Praise for
Supply Chain and Logistics Management Made Easy

"Paul Myerson's new book is a refreshing and a welcomed addition to the field, offering the reader a clear and easy-to-understand presentation of the key concepts and methods used in the field of supply chain management. His work is not only easy to understand but also comprehensive in coverage.

"I highly recommend it to university professors who want to incorporate it in their undergraduate and graduate courses in supply chain management. I have become a real fan of *Supply Chain and Logistics Management Made Easy*. Certainly, nothing in life is easy, but Paul Myerson's new book has made the field more attractive and popular."

—**Richard A. Lancioni**, Professor of Marketing and Supply Chain Management,
Fox School of Business & Management, Temple University

"Is it possible to take a discipline that involves millions of moving things, people, and processes and make it easy? Paul has taken the complex subject of supply chain and delivered a thorough and easy-to-understand review of all its elements. For the business student, the book provides a comprehensive view of the supply chain and serves as an effective introduction to the discipline and as an effective teaching tool. For the supply chain expert, this book is an excellent tool for reflection on all things supply chain. Each section brings back thoughts of the challenges the accomplished supply chain leader has faced. The book is an excellent resource for anyone in business who is looking to work in or currently works in supply chain management."

—**Gary MacNew**, Regional Vice President, Supply Chain Optimizers

"This is an excellent read for both students and professionals who are interested in gaining a better understanding of what supply chain and logistics is all about. It is an easy-to-understand handbook for anyone who has a need to better understand supply chain management or is responsible for helping their organization gain an advantage from their supply chain. Myerson's book should be on every manager's bookshelf for ready reference."

—**Robert J. Trent, Ph.D.**, Supply Chain Management Program Director,
Lehigh University

"Paul does a great job compacting supply chain management and logistics into one text. I wish I would have had this book when I was a logistics student 30+ years ago, but it's a great text and reference for me now, too. The SCM discipline is very wide and diverse now. This book captures all the elements. A complete professional reference. An easy read that teaches."

—**Andy Gillespie**, Director, Global Logistics, Ansell

"Practical, accessible, up-to-date, and covering today's best practices, *Supply Chain and Logistics Management Made Easy* is the ideal introduction to modern supply chain management for every manager, professional, and student."

—**Oliver Yao**, Associate Professor, Lehigh University

Supply Chain and Logistics Management Made Easy

SUPPLY CHAIN AND LOGISTICS MANAGEMENT MADE EASY

Methods and Applications for Planning, Operations, Integration, Control and Improvement, and Network Design

Paul A. Myerson
Professor of Practice in Supply Chain Management
Lehigh University

Publisher: Paul Boger
Editor-in-Chief: Amy Neidlinger
Executive Editor: Jeanne Glasser Levine
Operations Specialist: Jodi Kemper
Cover Designer: Alan Clements
Managing Editor: Kristy Hart
Project Editor: Andy Beaster
Copy Editor: Keith Cline
Proofreader: Sarah Kearns
Indexer: Erika Millen
Senior Compositor: Gloria Schurick
Manufacturing Buyer: Dan Uhrig

© 2015 by Pearson Education, Inc.

Old Tappan, New Jersey 07675

For information about buying this title in bulk quantities, or for special sales opportunities (which may include electronic versions; custom cover designs; and content particular to your business, training goals, marketing focus, or branding interests), please contact our corporate sales department at corpsales@pearsoned.com or (800) 382-3419.

For government sales inquiries, please contact governmentsales@pearsoned.com.

For questions about sales outside the U.S., please contact international@pearsoned.com.

Company and product names mentioned herein are the trademarks or registered trademarks of their respective owners.

© Miny – Fotolia.com
© volha – Fotolia.com

Printed in the United States of America

First Printing May 2015

ISBN-10: 0-13-399334-5
ISBN-13: 978-0-13-399334-9

Pearson Education LTD.
Pearson Education Australia PTY, Limited.
Pearson Education Singapore, Pte. Ltd.
Pearson Education Asia, Ltd.
Pearson Education Canada, Ltd.
Pearson Educación de Mexico, S.A. de C.V.
Pearson Education—Japan
Pearson Education Malaysia, Pte. Ltd.

Library of Congress Control Number: 2015932352

This book is dedicated to the memory of my father, Dr. Albert L. Myerson, the smartest person that I ever knew, who taught me the value of education and research.

I also appreciate the support of my wife, Lynne, and son, Andrew, without whose support and patience, this book would have taken a whole lot longer to write!

Contents

Part IV Supply Chain Integration and Collaboration 187

Chapter 12 Supply Chain Partners. 189

Chapter 13 Supply Chain Integration Through Collaborative Systems 199

Chapter 14 Supply Chain Technology 211

About the Author

Paul A. Myerson is a Professor of Practice in Supply Chain Management at Lehigh University and holds a B.S. in Business Logistics and an M.B.A. in Physical Distribution.

Professor Myerson has an extensive background as a Supply Chain and Logistics professional, consultant, and teacher.

Prior to joining the faculty at Lehigh, Professor Myerson has been a successful change catalyst for a variety of clients and organizations of all sizes, having over 30 years experience in Supply Chain and Logistics strategies, systems, and operations that have resulted in bottom-line improvements for companies such as General Electric, Unilever, and Church and Dwight (Arm & Hammer).

Professor Myerson created and has marketed a Supply Chain Planning software tool for Windows to a variety of companies worldwide since 1998.

He is the author of the books *Lean Supply Chain & Logistics* (McGraw-Hill, Copyright 2012) and *Lean Wholesale and Retail* (McGraw-Hill, Copyright 2014) as well as a Lean Supply Chain and Logistics Management simulation training game and training package (Enna.com, copyright 2012–13).

Professor Myerson also writes a column on Lean Supply Chain for *Inbound Logistics Magazine* and a blog for *Industry Week* magazine.

Part I

Supply Chain and Logistics Management: Overview

1

Introduction

In the early 1980s, U.S. companies dramatically increased the outsourcing of manufacturing, raw materials, components, and services to foreign countries. Around that time, the term *supply chain* was coined to recognize the increased importance of a variety of business disciplines that were now much more challenging to manage as a result of the new global economy. Prior to that, functions such as purchasing, transportation, warehousing, and so on were isolated and at fairly low levels in organizations.

Since that time, we've seen the creation of the Internet and various business technologies such as enterprise resource planning (ERP) systems, advanced planning systems (APS), and radio frequency ID (RFID), to name a few, which have helped to speed up the flow of information and product lifecycles as well as increasing the need for better communication, collaboration and visibility.

Today, logistics alone accounts for more than 9.5% of U.S. gross domestic product (GDP). Over $1.3 trillion is spent on transportation, inventory, and related logistics activities. The concept of the supply chain has now risen in importance to the extent that commercials on TV extol the virtues of logistics (for example, UPS "I Love Logistics" commercials) to the point where it is now part of the common lexicon and very mainstream. As a result, most universities now offer supply chain and logistics courses, if not majors, and most organizations have a vice president of supply chain and logistics management (or similar title).

However, beyond supply chain and logistics employees, not many in business or the public fully understand the role and importance that the supply chain plays in gaining and maintaining a competitive advantage in today's world.

We are at the point today where most people are familiar with the terms *supply chain* and *logistics* but don't really know that much about them. In this book, we not only define the supply chain but also offer insight into its various components, tools, and technology to help improve your understanding so that you can use it as a competitive tool in your business.

Because supply chain and logistics costs can range from 50% to 70% of a company's sales (with trillions spent on it worldwide), organizations of all sizes both perform and are interested in this function. Therefore, understanding and implementing an efficient supply chain strategy can prove critical to both an employee's and a company's success.

Supply Chain Defined

The first thing we need to do is get some definitions out of the way. The terms *supply chain* and *supply chain management* (SCM) should be separately defined because they are sometimes (mistakenly) used interchangeably.

The supply chain itself is a system of organizations, people, activities, information, and resources involved in the planning, moving, or storage of a product or service from supplier to customer (actually more like a "web" than a "chain"). Supply chain activities transform natural resources, raw materials, and components into a finished product that is delivered to the end customer. For example, I once heard a major paper goods manufacturer describe their supply chain for toilet paper as ranging from "stump to rump."

In contrast, *supply chain management,* as defined by the Council of Supply Chain Management Professionals (CSCMP), "encompasses the planning and management of all activities involved in sourcing, procurement, conversion, and logistics management. It also includes the crucial components of coordination and collaboration with channel partners, which can be suppliers, intermediaries, third-party service providers, and customers."

In essence, supply chain management integrates supply and demand management within and across companies and typically "includes all of the logistics management activities noted above, as well as manufacturing operations, and it drives coordination of processes and activities with and across marketing, sales, product design, finance and information technology" (Council of Supply Chain Management Professionals [CSCMP], 2014).

Some people take a narrower view of supply chain, and in many cases, they think of it as focused more on the supply end (that is, purchasing), and so ignore the logistics side (as defined as the part of the supply chain that plans, implements, and controls the efficient movement and storage of goods, services, and information from the point of use or consumption to meet customer requirements). In other cases, many just assume that logistics is included but don't state it. Still others, while including both areas above, ignore the planning aspects of supply chain. Personally, I tend to refer to the field as *supply chain and logistics management* to make clear what is included.

As you will see in this book, it is important to understand the similarities and differences between more functional areas like logistics, which includes transportation and distribution, versus the broader concept of SCM, which is cross-functional and cross-organizational. This can have a major impact on decision making, structure, and staffing in an organization, so it needs to be understood and examined carefully.

Depending on one's view, some of the functions listed here may be included within the supply chain and logistics organization:

- **Procurement:** The acquisition of goods or services from an outside external source
- **Demand forecasting:** Estimating the quantity of a product or service that customers will purchase
- **Customer service and order management:** Tasks associated with fulfilling an order for goods or services placed by a customer
- **Inventory:** Planning and management
- **Transportation:** For hire and private
- **Warehousing:** Public and private
- **Materials handling and packaging:** Movement, protection, storage, and control of materials and products using manual, semi-automated, and automated equipment
- **Facility network:** Location decision in an organization's supply chain network

Supply chain management is also intertwined with operations management, which consists of activities that create value by transforming inputs (that is, raw materials) into outputs (that is, goods and services). Both activities support the manufacturing process.

SCOR Model

Another way to view the supply chain is through the SCOR model, which was developed by the Supply Chain Council (SCC) (2014) to teach, understand, and manage supply chains. It is a model to both define and measure the performance of an organization's supply chain.

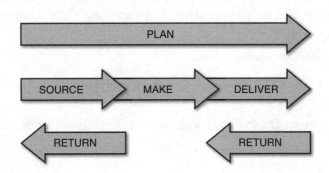

Figure 1.1 SCOR model

The SCOR model is organized around the five major management processes (see Figure 1.1):

- **Plan:** Alignment of resources to demand
- **Make:** Conversion or value-added activities within a supply chain operation
- **Source:** Buying or acquiring materials or services

- **Deliver:** All customer interaction, from receiving order to final delivery and installation
- **Return:** All processes that reverse material or service flows from the customer backward through the supply chain

This provides a broad definition for the supply chain, which highlights its importance to the organization and how it helps create metrics to measure performance.

SCOR Metrics

To this aim, the SCOR model is also a hierarchical framework that combines business activities, metrics, and practices that can be looked at from a high or very detailed level.

The levels, from broadest to narrowest, are defined as follows:

Level 1: Scope: Defines business lines, business strategy and complete supply chains.

Level 2: Configuration: Defines specific planning models such as "make to order" (MTO) or "make to stock" (MTS), which are basically process strategies.

Level 3: Activity: Specifies tasks within the supply chain, describing what people actually do.

Level 4: Workflow: Includes best practices, job details, or workflow of an activity.

Level 5: Transaction: Specific detail transactions to perform a job step.

All SCOR metrics have five key strategic performance attributes. A performance attribute is a group of metrics used to express a strategy. An attribute itself cannot be measured; it is used to set strategic direction.

The five strategic attributes are as follows:

- **Reliability:** The ability to deliver, on time, complete, in the right condition, packaging, and documentation to the right customer
- **Responsiveness:** The speed at which products and services are provided
- **Agility:** The ability to change (the supply chain) to support changing (market) conditions
- **Cost:** The cost associated with operating the supply chain
- **Assets:** The effectiveness in managing assets in support of demand satisfaction

The SCOR model contains more than 150 key indicators, such as inventory days of supply and forecast accuracy, that measure the performance of supply chain operations and are grouped within the previously listed strategic attribute categories.

Once the performance of supply chain operations has been measured and performance gaps identified, they are *benchmarked* against industry best practices to target improvement, as discussed in more detail later in this book.

An Integrated, Value-Added Supply Chain

The goal for today's supply chain is to achieve integration through collaboration to achieve visibility downstream toward the customer and upstream to suppliers. In a way, many of today's companies have been able to "substitute information for inventory" to achieve efficiencies. The days of having "islands of automation," which may optimize your organization's supply chain at the cost of someone else's (for example, your supplier), are over.

As you will see throughout this book, the concepts of teamwork and critical thinking aided by technology enable organizations to work with other functions internally and with other members of their supply chain, including customers, suppliers, and partners, to achieve new levels of efficiency and to use their supply chain to achieve a competitive advantage that focuses on adding value to the customer as opposed to just being a cost center within the organization.

The Value Chain

The Value Chain model, originated by Michael Porter, shows the value-creating activities of an organization, which as you can see in Figure 1.2 relies heavily on supply chain functions.

In a value chain, each of a firm's internal activities listed after the figure adds incremental value to the final product or service by transforming inputs to outputs.

| Inbound Logistics | Operations | Outbound Logistics | Sales and Marketing | Service |

Figure 1.2 The value chain

- **Inbound logistics:** Receiving, warehousing, and inventory control of input materials
- **Operations:** Transforming inputs into the final product or service to create value
- **Outbound logistics:** Actions that get the final product to the customer, including warehousing and order fulfillment

- **Marketing and sales:** Activities related to buyers purchasing the product, including advertising, pricing, distribution channel selection, and the like
- **Service:** Activities that maintain and improve a product's value, including customer support, repair, warranty service, and the like

Support activities identified by Porter can also add value to an organization:

- **Procurement:** Purchasing raw materials and other inputs that are used in value-creating activities
- **Technology development:** Research and development, process automation, and similar activities that support value chain activities
- **Human resource management:** Recruiting, training, development, and compensation of employees
- **Firm infrastructure:** Finance, legal, quality control, and so on

Porter recommended value chain analysis to investigate areas that represent potential strengths that can be used to achieve a competitive advantage. As you can see, the supply chain adds value in a variety of ways, so it should be a critical area of focus (Porter, 1985).

We investigate ways to identify value-added and non-value-added activities (which should be reduced or eliminated) in a supply chain later in this book using a Lean methodology and tools.

Leveraging the Supply Chain

Because supply chain costs represent a significant portion of a company's sales, it isn't difficult to see why there is such a focus on it. This results in a "leveraging" effect, as any dollar saved on supply chain contributes as the same to the bottom line as a much larger and often unattainable increase in sales (will vary based on an individual company's profit margin).

Table 1.1 illustrates this through an example of a business that is evaluating two strategic options: 1) reduce its supply chain costs by approximately 6.5% through more effective negotiations with a vendor, or 2) increase sales by 25% (which will most assuredly also add to sales and marketing costs). You can see the leveraging effect of the supply chain as the relatively small cost decrease contributes as much to the bottom line as the 25% sales increase (which is pretty difficult to accomplish in any economy).

Table 1.1 Supply Chain Leveraging Effect

	Current	Supply Chain Improvement Option	Sales Increase Option
Sales	$1,000,000	$1,000,000	**$1,250,000**
Cost of material	$650,000 (65%)	**$600,000 (60%)**	$812,500 (65%)
Production costs	$150,000 (15%)	$150,000 (15%)	$187,500 (15%)
Fixed costs	$100,000 (10%)	$100,000 (15%)	$100,000 (8%)
Profit	$100,000 (10%)	$150,000 (15%)	$150,000 (15%)

The supply chain cost reduction in this example has impressive results, but you have to keep in mind that "you can't get blood from a stone." That is where Lean techniques, which are discussed later, can have a significant impact. Through Lean, a team-based form of continuous improvement that focuses on the identification and elimination of waste, we can create a "paradigm shift" that can make process (and cost) improvements that were previously thought impossible.

Supply Chain Strategy for a Competitive Advantage

Historically, supply chain and logistics functions were viewed primarily as cost centers to be controlled. It is only in the past 20 years or so that it has become clear that it can be used for a competitive advantage as well.

To accomplish this, an organization should establish competitive priorities that their supply chain must have to satisfy internal and external customers. They should then link the selected competitive priorities to their supply chain and logistics processes.

Krajewski, Ritzman, and Malhotra (2013) suggest breaking an organization's competitive priorities into cost, quality, time, and flexibility capability groups:

- **Cost strategy:** Focuses on delivering a product or service to the customer at the lowest possible cost without sacrificing quality. Walmart has been the low-cost leader in retail by operating an efficient supply chain.

- **Time strategy:** This strategy can be in terms of speed of delivery, response time, or even product development time. Dell has been a prime example of a manufacturer that has excelled at response time by assembling, testing, and shipping computers in as little as a few days. FedEx is known for fast, on-time deliveries of small packages.

- **Quality strategy:** Consistent, high-quality goods or services require a reliable, safe supply chain to deliver on this promise. If Sony had an inferior supply chain with high damage levels, it wouldn't matter to the customer that their electronics are of the highest quality.

- **Flexibility strategy:** Can come in various forms such as volume, variety, and customization. Many of today's e-commerce businesses, such as Amazon, offer a great deal of flexibility in many of these categories.

In many cases, an organization may focus on more than one of these strategies, and even when focusing on one, it doesn't mean that they will offer subpar performance on the others (just not "best in class" perhaps).

Segmenting the Supply Chain

Today's use of "omni-channel marketing," which is an integrated approach of selling to consumers through multiple distribution channels (that is, brick-and-mortar, mobile Internet devices, computers, television, radio, direct mail, catalog, and so on) has created the need to handle multiple channels with separate warehouse picking operations, often replenished from a common inventory in a single facility.

This can lead many companies such as Dell Computer to segment their entire supply chains, whereby different channels and products are served through different supply chain processes. The ultimate goal is to determine the best supply chain processes and policies for individual customers and products that also maximize customer service and company profitability.

The rationale for this, according to an Ernst & Young white paper titled "Supply Chain Segmentation," (2014) is that the "business environment is getting increasingly complex, especially for technology companies dealing with rapid innovation, globalization, and a growing number of business partners, business models, and differences in expectations from different markets and customers."

E&Y suggest five ways to consider segmentation:

- Product complexity based
- Supply chain risk based
- Manufacturing process and technology based
- Customer service needs based
- Market driven

The idea is that a "one size fits all" strategy will not usually work in today's environment.

They suggest that while senior sponsorship is required for successful supply chain segmentation, you also need cross-functional support from multiple organizational disciplines. The team must provide supporting policies, segment-level processes, and IT infrastructure to both automate the processes and provide metrics.

In Dell's case, over the past few years, they have expanded beyond their direct to customer model to a "multichannel, segmented model, with different policies for serving consumers, corporate customers, distributors, and retailers. Through this transformation, Dell has saved US $1.5 billion in operational costs" (Thomas, 2012).

The Global Supply Chain and Technology

Suffice it to say, the concept of "global" supply chain management (GSCM) is primarily a result of the globalization of business in general. As businesses search globally for sources of lower-cost materials and labor, someone has to manage these complex and intricate operations.

The combination of globalization and emerging technologies is continuously changing the supply chain. Products that were once made domestically, such as apparel and computers, are now designed, assembled, and marketed worldwide by a conglomeration of organizations.

As a result, there are many risks (disruptions, natural disasters, domestic job loss, and so on) and challenges (short product lifecycles, erratic demand, and so on) that are inherent to the process. To this end, a roundtable at a Dartmouth University Roundtable identified five major issues and challenges ahead (Johnson, 2006):

- **Globalization and outsourcing:** Including the impact of China and India on supply chain structure and coordination
- **New information technologies:** Such as radio frequency identification (RFID; a data collection technology that uses electronic tags for storing data) and tools that enable enterprise integration and collaboration
- **Economic forces:** Within and between supply chains, from consumer pricing to supplier contract negotiation
- **Risk management:** Includes risks rising from supply chain complexity and from security threats
- **Product lifecycle management:** Including post-sale service and product recovery

We discuss the impact of global operations and various forms of technology used today in supply chain management later in this book.

For now, we will look back to get a little historical perspective on the topic of supply chain and logistics management.

2

Understanding the Supply Chain

Over the past 75 years, supply chain and logistics management has evolved from being a kind of "back water" cost-focused function (the term *supply chain* didn't even come about until the 1980s) to where it is today, a critical part of an organization's global growth strategy. It has become a strategic competitive tool to increase revenue and value to the customer, not just reduce costs.

Once one has a broader understanding of the topic as covered in this chapter, it's not hard to see why it has risen in importance in recent times.

Historical Perspective

Until after World War II, logistics was thought of in military terms for the most part as the link that supplied troops with rations, weapons, and equipment. Up to that point, logistics was fragmented within business organizations, primarily focusing on transportation and purchasing. In educational institutions, there were no integrated programs. Instead, individual courses were offered in transportation and purchasing.

After World War II, as businesses began to understand the relationships and tradeoffs involved such as inventory costs versus transportation costs, which are discussed later, logistics gained an important place in the business world as well.

In the 1960s, physical distribution, a more integrated concept that included activities such as transportation, inventory control, warehousing, and facility location had become an area of study and practice in education and industry. Physical distribution involved the coordination of more than one activity associated with supplying product to the marketplace (that is, more focused on the *outbound* side of manufacturing).

In the mid-1960s, the scope of physical distribution was expanded to include the supply side, including inbound transportation and warehousing, and was referred to as *business logistics*. In many cases, purchasing was not included and went under the heading of *materials management* or *procurement*.

In the early 1980s, as American manufacturing had been hammered by overseas competitors for over a decade and began actively outsourcing materials, labor, and manufacturing overseas, the term *supply chain management* (SCM) entered the common business lexicon. It defined both the new, complex global world we now live and do business in, as well as an understanding of the integration and importance of all activities involved in sourcing and procurement, conversion, and logistics management. This includes the coordination and collaboration with channel partners, which can be suppliers, intermediaries, third-party service providers, and customers.

In contrast to the past, where physical distribution, logistics, purchasing, and so on were all fragmented, many of today's organizations feature in integrated supply chain organization in most cases led by a senior-level executive (see Figure 2.1).

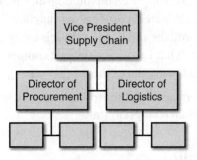

Figure 2.1 Supply chain organizational chart

Technology has helped to drive the concept of an integrated supply chain starting with the development of electronic data interchange (EDI) systems as a standardized format for the electronic transfer of data between business enterprises (which really took off in the 1980s), as well as the introduction of off-the-shelf enterprise resource planning (ERP) software systems, which featured integrated core business processes in a common database. Furthering this into the 21st century has been the expansion of Internet-based collaborative systems.

This supply chain evolution has resulted in both increasing value added and cost reductions through integration and collaboration.

Value as a Utility

Utility refers to the value or benefit a customer receives from a purchase. There are four basic types of utility: form, place, time, and possession. More recently, the utilities of information and service have been added. SCM contributes to all of these utilities, because it's all about having the *right product, at the right place and price, at the right time.*

Here are descriptions of each utility:

- **Form utility:** Performed by the manufacturers (as well as third-party logistics companies, or 3PLs), which perform value-added activities such as kitting and display assembly, making the products useful.
- **Time utility:** Having products available when needed.
- **Place utility:** Having items available where people want them.
- **Possession utility:** Transfer ownership to the customer as easily as possible, including the extension of credit.
- **Information utility:** Opening two-way flows between parties (that is, customer and manufacturer).
- **Service utility:** Providing fast, friendly service during and after the sale and teaching customers how to best use products. This is becoming one of the most important utilities offered by retailers, which in many ways are part of the supply chain.

Organizational and Supply Chain Strategy

If an organization can identify what adds value to their customers and deliver it successfully, they will have established a competitive advantage, which in essence is the purpose of a strategic plan.

Mission Statement

To do so, you must first establish a broad mission statement, supported by specific objectives for your business. A mission statement is a company's purpose or *reason for being* and should guide the actions of the organization, lay out its overall goal, providing a path, and guiding decision making.

It doesn't have to be lengthy, but should be well thought out and touch on the following concepts:

- **Customers:** Who are our customers?
- **Products or services:** Major products or services.
- **Markets:** Where do we compete?
- **Technology:** What is our basic technology?
- **Future survival, growth, and profitability:** Our commitment toward economic objectives.
- **Philosophy:** The basic beliefs, core values, aspirations, and philosophical priorities of the firm.

- **Self-concept:** Identify the firm's major strengths and competitive advantages.
- **Public image:** What is our public image?
- **Concern for employees:** Our attitude toward employees.

Objectives

The *mission* is a broad statement, but should then lead to specific objectives with measurable targets a firm can use to evaluate the extent to which it achieves its mission.

In a typical medium- to large-size organization, individual functions/departments may have their own mission statements, but most at least have goals and objectives that tie to the company's overall mission statement and objectives (see Table 2.1).

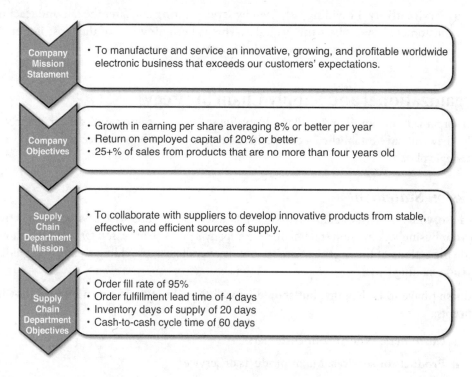

Company Mission Statement
- To manufacture and service an innovative, growing, and profitable worldwide electronic business that exceeds our customers' expectations.

Company Objectives
- Growth in earning per share averaging 8% or better per year
- Return on employed capital of 20% or better
- 25+% of sales from products that are no more than four years old

Supply Chain Department Mission
- To collaborate with suppliers to develop innovative products from stable, effective, and efficient sources of supply.

Supply Chain Department Objectives
- Order fill rate of 95%
- Order fulfillment lead time of 4 days
- Inventory days of supply of 20 days
- Cash-to-cash cycle time of 60 days

Table 2.1 Company and Supply Chain Mission Statements and Objectives

SWOT Analysis

After specifying the objectives for a business, an organization should perform a *SWOT* analysis (strengths, weaknesses, opportunities, and threats) to determine strategic choices for the organization to establish a competitive advantage.

The components of a SWOT analysis are as follows:

- **Strengths:** Resources and capabilities that can be used as a basis for developing a competitive advantage.
- **Weaknesses:** Characteristics that place the business or project at a disadvantage relative to other businesses.
- **Opportunities:** External environmental analysis may reveal certain new opportunities for profit and growth.
- **Threats:** Changes in the external environmental may also present threats to the firm.

Using the SWOT framework, you can start to develop a strategy that helps you distinguish your organization from your competitors so that you can compete successfully in your markets.

Strategic Choices

Strategic choices will be made based on the results of the SWOT analysis and can fall into the competitive priority categories described in Chapter 1, "Introduction," of cost, quality, time, or flexibility.

The supply chain must then be managed to support these strategies.

Supply Chain Strategy Elements and Drivers

Hernàn David Perez, paraphrasing Michael Porter, suggests that "supply chain strategy defines the connection and combination of activities and functions throughout the value chain, in order to fulfill the business value proposal to customers in a marketplace" (Perez, 2013).

As a result, he describes how the supply chain strategy is driven by the interrelation among four main elements (see Figure 2.2):

- Industry framework (that is, the marketplace)
- Company's value proposition to the customer via its competitive positioning
- Managerial focus (relationship between supply chain processes and business strategy)
- Internal (supply chain) processes

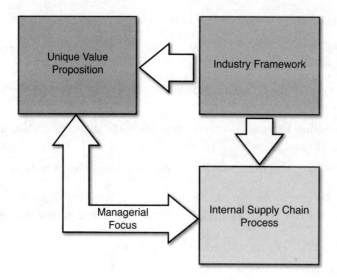

Figure 2.2 Four main elements of supply chain strategy

Industry Framework

Developing an industry framework involves identifying the interaction of suppliers, customers, technological developments, and economic factors that may impact competition.

Four drivers can impact supply chain design:

- **Demand variation:** This can be a wide array of manufacturing and supply chain costs and is therefore a major driver of efficiency and ultimately cost.

- **Market mediation costs:** Costs incurred when supply doesn't match demand, often resulting in either lost sales or higher than needed supply chain costs and excess inventory.

- **Product lifecycle:** Advances in technology as well as consumer trends have reduced the time to bring an item to market as well as its useful life. Affects demand variability as well as marketing and supply chain costs.

- **Relevance of the cost of assets to total cost:** Largely affects businesses requiring a high utilization rate to remain profitable (for example, chemical industry). This encourages a *push* mentality to gain high utilization of assets but can result in higher than inventory costs and lower service levels. Industries that have lower cost assets can focus on being more responsive.

Unique Value Proposal

The value proposition offered to a company's customer is best understood after the establishment of the competitive priority strategy that the organization has selected in terms of its supply chain. As part of the organization's strategy, they need to determine what it takes to win business and incorporate that into their value proposition, thereby understanding and incorporating the required key drivers into their supply chain to that ultimately the required value is delivered to the customer.

Managerial Focus

To be successful, an organization must make sure that its supply chain is linked and aligned with its competitive priorities. This can only be accomplished via its decision-making process and management focus.

It can be very easy for management to only focus on efficiency-oriented performance measurements at the expense of the competitive priorities set by the company. A misalignment can result with the supply chain being *suboptimized* by attaining local cost efficiencies at the cost of the value proposition offered to the customer.

Internal Processes

Internal processes must be connected and aligned properly. Thinking in terms of the Supply Chain Operations Reference (SCOR) model processes of source, make, and deliver can help to make sure this occurs. It is of critical importance to determine the appropriate *decoupling point* (that is, where a product takes on unique characteristics or specifications). This goes in to determining which parts of your internal processes are *push* (that is, high asset utilization rate; just before the decoupling point) versus *pull* (that is, workload driven by customer demand).

Supply Chain Strategy Methodology

So how might one go about actually establishing a departmental or functional supply chain strategy for your organization? Paul Dittman, in his book *Supply Chain Transformation* (Dittman, 2012), suggests using nine steps when developing a supply chain strategy. I have modified those slightly for purposes of clarity and results:

1. **Start with customers' current and future needs.** Customer value is the customer-perceived benefits gained from a product/service compared to the cost of purchase. Delivering customer value is critical to a business, as mentioned earlier.

   However, delivering financial value to your shareholders is also important, and is reflected in various business performance measures such as profit and market growth.

Supply chain strategy should target to deliver customer value *while at the same time* meet shareholder needs by enabling reliable supply and logistics service, low inventory cost, and short cash-to-cash cycle times.

Using the SCOR model processes of plan, source, make, deliver, and return can be a great way to make sure that customer and shareholder value are in alignment.

(Steps 2–6 that follow involve using a kind of SWOT analysis for your supply chain organization.)

2. **Assess current supply chain capabilities relative to best in class.** This can be accomplished through observation, interviews, data gathering, and benchmarking your organization against industry best-in-class performance.

   Based on your organization's overall strategy, some metrics and measurement may be more important than others. For example, if you have a time-based strategy, speed of delivery may be important, whereas cost of delivery (relatively speaking) may not be as important in terms of achieving best-in-class status.

   Developing a *gap analysis* of your current versus ideal future state based on this analysis can contribute to a clear and easy-to-follow roadmap.

3. **Evaluate supply chain game changers.** It is important to scan the environment on a regular basis to see what trends may impact customers and the supply chain. Examples include supply chain collaboration, visibility, sustainability, Lean Six Sigma, and so on (some of which we discuss later in this chapter in the "Supply Chain Opportunities and Challenges" section).

4. **Analyze the competition.** As the saying goes, "Keep your friends close but your enemies closer." Competitive analysis is probably not done often enough in terms of an organization's supply chain. If you plan on using your supply chain to achieve a competitive advantage, this is a *must do.*

   You may evaluate how integrated and responsive the competition's supply chain is when compared to yours and if they offer value-added services, such as the following:

   - Product customization and testing
   - Kitting
   - Bundling/unbundling
   - Light assembly
   - Packaging, repackaging, and reboxing
   - Labeling
   - Sorting and recycling
   - Reverse logistics and returns management
   - Environmental impact reporting and management

5. **Survey technology.** Don't just identifying what is new or being developed, but what is a good fit (functionally and financially) for your company. Sometimes it is better to not be on the bleeding edge when it comes to technology. When I was at Uniliver, they had spent hundreds of thousands of dollars on artificial intelligence technology to deploy finished goods inventory to their distribution centers. It never really went anywhere (at least to my knowledge), for a variety of reasons.

6. **Deal with supply chain risk.** Risk management needs to be part of the strategy document. External risks, which are out of your control, can be driven by events either upstream or downstream in the supply chain. Here are the main types of external risks:

   - *Demand risks:* Can be caused by unpredictable customer or end-customer demand.

   - *Supply risks:* These types of risks are caused by interruptions to the flow of product for raw material or components, within your supply chain. For example, if you are utilizing a just-in-time (JIT) strategy for a critical part or component, you need to think long and hard as to what risks are involved, because you do not want to risk shutting down a production line due to a critical part that you have sole sourced suddenly becoming unavailable.

   - *Environmental risks:* Come from outside the supply chain. These risks usually relate to economic, social, governmental, and climate factors, and include the threat of terrorism.

   - *Business risks:* Can include a supplier's financial or management stability or purchase and the sale of supplier companies.

   - *Physical plant risks:* This risk can be caused by the condition of a supplier's physical facility and regulatory compliance.

   Now that you have identified what adds value to your customers while making sure it is aligned with shareholder needs, as well as identified possible current and future performance gaps in your supply chain, a road map for future success can be developed.

7. **Develop new supply chain capability requirements and create a plan to get there.** One way to determine these requirements was formulated by Hau Lee, who concluded that supply chains that offer best value to the customer differ from typical supply chains in how they approach three issues that are closely tied to strategic supply chain management (Lee, 2004):

   - *Agility:* The supply chain's relative capacity to act rapidly in response to dramatic changes in supply and demand

   - *Adaptability:* Refers to a willingness and capacity to reshape supply chains when necessary

- *Alignment:* Refers to creating consistency in the interests of all participants in a supply chain

In Lee's model, these three A's can be used to service an organization's competitive priorities, as discussed earlier (see Figure 2.3).

Issue	Typical Supply Chain	Best Value Supply Chain	Example Company
Agility	Moderate capacity to react to changes	Good capacity to anticipate and react to changes	Raytheon Technical Services Company locates an executive office nearby key customers
Adaptability	Focus on efficiency through the use of discrete supply chains	Maintain overlapping supply chains to ensure customer service	Computer Sciences Corporation positions some inventory close to customer locations while other items are warehoused centrally
Alignment	Supply chain members sometimes forced to choose between their own interest and the chain's interest	A rising tide lifts all boats; the interests of supply chain members are consistent with each other	When a supplier's suggestion saves R.R. Donnelly money, the firm splits the savings with the customer.

Figure 2.3 Typical to best value supply chains

8. **Evaluate current supply chain organizational structure, people, and metrics.** There is no one-size-fits-all approach for creating an organization. Traditional supply chain organizations are functionally oriented. In the 1980s and 1990s, companies started to transition to structures that grouped some core supply chain functions within one department. From around 2000 onward, the philosophy of the supply chain as an end-to-end process took hold more often than not with a director or vice president of supply chain overseeing the operation. This also requires giving that manager a set of cross-functional performance objectives (and metrics) and the resources they require to meet these objectives

This type of organization requires an evaluation of existing capabilities and identification of any gaps between currently available skills and those needed to support this end-to-end strategy.

9. **Develop a business case and get buy-in.** Of course, any type of change typically has to be approved by management. To do this, you need to develop a business case, because whenever resources such as money or effort are utilized, they should be in support of a specific business need. An example of a business case for a new supply chain strategy might state that "improvement initiatives outlined in the plan will have a broad impact throughout the entire company, increasing efficiency and aligning business activities across all lines of business. Different aspects of the enterprise can now coordinate their procurement efforts and material flows to increase operating efficiency, and take advantage of their combined buying power to negotiate better prices and contract terms."

Supply Chain Opportunities and Challenges

When considering a supply chain strategy, it is important to be aware of current opportunities and challenges.

We live in volatile times. There are many external threats to our supply chain now and in the future. A white paper by Hitachi Consulting identified "Six Key Trends Changing Supply Chain Management Today" (Hitachi, 2009):

- **Demand planning:** Companies are moving more toward a "make what you sell" demand or pull-driven process and are trying to influence and manage demand better. Cross-functional processes such as sales & operations planning (S&OP), which are discussed later in this text, help to improve awareness, coordination, and accuracy of demand estimates to ultimately improve customer service while reducing costs.

  Organizations that do not wrap their minds around this will continue to struggle with meeting ever-increasing volatile demand, which in many cases is caused by incentives and promotions resulting in manufacturing (and purchasing) producing larger-than-needed lot sizes, which are *pushed* through the supply chain, causing inefficiencies throughout. This is known as the *bullwhip effect* (see Figure 2.4), which describes the magnification (especially on inventory, operational costs, and customer service) that occurs when orders move up the supply chain. This can be caused by a variety of things such as forecast errors, large lot sizes, long setups, panic ordering, variance in lead times, and so on.

  We discuss ways to combat the bullwhip effect throughout this book using techniques and tools such as Lean to reduce production and distribution *pushing* of large batches, EDI (electronic data interchange) to avoid batching of orders, collaboration programs such as VMI (vendor-managed inventory) and CPFR (collaborative planning, forecasting, and replenishment) to collaborate with customers and suppliers to share information, everyday low pricing (EDLP), and so on.

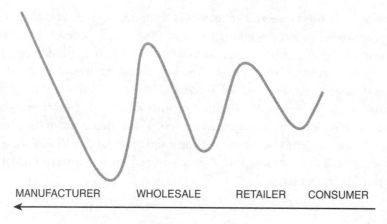

MANUFACTURER   WHOLESALE   RETAILER   CONSUMER

Figure 2.4 The bullwhip effect

- **Globalization:** No area of business is more impacted by globalization than the supply chain. The benefits to globalization include access to more markets, a larger supplier pool, a greater selection of employees, and so on. On the downside are the various risks mentioned earlier.

  Supply chain network design is important to managing the changes brought about by globalization and can optimize the number, location, size, and type of facilities and flow of materials throughout the network.

- **Increased competition and price pressures:** The commoditization of many products has forced businesses to find better ways to distinguish themselves. They now look to the supply chain to reduce cost and add value to the customer through both the product and service.

  Cost improvements can be found through the following:

  - Sales and operations planning
  - Transportation/distribution management
  - Improved product lifecycle management
  - Improved strategic sourcing and procurement

  Value-added service can be provided through the following:

  - *Vendor-managed inventory (VMI):* Buyer of a product provides information to a vendor, and the supplier takes full responsibility for maintaining an agreed inventory of the material, usually at the buyer's consumption location.
  - *Radio frequency identification (RFID):* The wireless use of radio frequency to transfer data, to identify and track tags attached to objects.

- Labeling and packaging
- Drop shipping
- Collaboration

- **Outsourcing:** The supply chain and logistics functions are always a good candidate for outsourcing because they may not be a core competency for an organization. There is, of course, the tradeoff of risk and reward, which requires good supply chain network design integration with the outsourcing partner in the information chain, control mechanisms to monitor the various components of the supply chain, and information systems that connect and coordinate the entire supply chain.

- **Shortened and more complex product lifecycles:** Today there is increasing pressure to develop and introduce new and innovative products quickly. To do this, companies have worked on improving their product lifecycle management (PLM) processes. Benefits of PLM to the supply chain include processes and technology to design products that can share common operations, components, or materials with other products. This can reduce the risk of obsolescence and reduce costs when purchasing key materials. A formalized PLM process can also help to coordinate marketing, engineering, sales, and procurement and develop sales forecasts to plan products that are in a company's pipeline

- **Closer integration and collaboration with supplier:** Supply chain collaboration is more than just connecting information systems and now extends to fully integrating business processes and organization structures across companies that make up the entire value chain.

  S&OP processes now extend to an organization's external supply chain partners to include demand information, such as customer forecasts, and supply information, such as supplier inventories and capacities.

As the supply chain has become more global and complex in nature, technology has advanced to help manage the process and comes under the heading of systems such as Enterprise Resource Planning (ERP), Supply Chain Management (SCM), and supply chain planning (SCP) systems. These systems enable processes such as the following:

- Network and inventory optimization
- Logistics optimization
- Product lifecycle management
- Sales and operations planning
- Procurement
- Manufacturing optimization

- Warehouse operations
- Business intelligence

I would also add the trend of companies that have started applying Lean and Six Sigma concepts to the extended supply chain. These are both team-based continuous-improvement processes. Lean helps to identify and eliminate waste throughout an organization, and Six Sigma reduces variation in individual processes. Because they are somewhat complementary, they have now, in many cases, been combined as *Lean Six Sigma*.

There is also the challenge of sustainability as resources have become increasingly constrained due to the global economy and climate change, which has led to governmental regulations that attempt to minimize damage to the environment.

All of these issues are covered in more detail in different sections of this book.

Supply Chain Talent Pipeline

The growth of the importance of the supply chain and logistics field also requires good talent that has a fundamental understanding of both supply chain and logistics concepts and the proper tools and training to analyze and improve it.

Many universities offer great supply chain programs, such as Lehigh University (where I teach), but there are, according to a study, "a number of key emerging trends that individually create tension and potential disruptions in the supply chain talent pool. Either of those on their own can create challenges for a supply chain organization similar to a hurricane or a severe winter gale. At the same time, like the (movie) *Perfect Storm*, there is the prospect of these trends colliding to create a supply chain talent 'perfect storm'" (Ruamsook & Craighead, 2014).

The authors of the study mention that demand for supply chain talent is projected to continue to rise, while the talent gap will become greater as baby boomers start retiring. That, combined with the increased need for technical skills in a more complex global economy and a possible shortage in supply chain university faculty in the coming years, may be the impetus for a *perfect storm* of sorts.

We all know how accurate the weather forecasts have been lately, so no need to panic in my opinion. However, it is important to think strategically to avoid potential obstacles like this.

The paper also points out the need to plan ahead by focusing on the employee value proposition (that is, opportunity, rewards, and so on), making sure that you hire people with the right *core competencies*, focusing on retention methods, investing in talent and leadership development, and helping to create a *talent pipeline* by working with high schools and universities to develop the talent.

Supply chain educators try to do their part by encouraging business students to consider supply chain as a major (or minor), while giving them a great fundamental understanding of supply chain and logistics management practical concepts and applications (and actual experience through internships and team projects). They also help them to develop the skills and abilities necessary to understand and use the various tools and technology available today to manage and improve the supply chain process in this complex global economy.

The authors of the aforementioned article emphasize the need for a tighter industry-academic collaboration to help avoid this perfect storm.

It is important that supply chain organizations prepare properly in this regard (and in general) and take the long view to plan ahead so that they will be able to ride out this or any storm successfully.

Career Opportunities in Supply Chain and Logistics Management

The supply chain talent pipeline can be employed in a variety of areas, many which might not be totally obvious to those interested in careers in this field.

Depending on one's interests, talents, and aptitude, careers can be found in the corporate environment, field operations, and even sales in functional areas such as forecasting, production and deployment planning, manufacturing, transportation, distribution, procurement, and technology.

When thinking about a supply chain career path, it helps to consider preferences and strengths with regard to mathematics, critical thinking, social, travel, work environment, and so on, because this may steer you down one path versus another.

Table 2.2 lists a number of jobs in the field.

Growing Demand

According to Bloomberg Business Week, SCM "job openings, comfortable salaries, and the prospect for advancement have caused the academic community to take notice, with more students majoring in the subject and more programs offering courses and concentrations in it" (Taylor, 2011).

For example, Lehigh University's College of Business and Economics has recently reported the most undergraduate SCM majors in the program's 10-year history.

The article goes on to report that SCM majors and MBAs are in high demand and that the average entry-level professional supply management salary is about $49,500 and the average salary of those with 5 or fewer years of experience is $83,689.

Table 2.2 Supply Chain and Logistics Job Sampling

Supply Chain	Forecasting / Planning	Purchasing / Procurement	Logistics	Operations	Inventory Management	Transportation	Warehousing and Distribution Shift Manager	Customer Service
Supply chain analyst	Expeditor	Assistant buyer / purchasing	Logistics analyst, Operations manager	Expeditor	Transportation coordinator	Warehouse supervisor	Customer service assistant	Supply chain analyst
Supply chain coordinator	Materials planner	Buyer, Senior buyer	Logistics coordinator	Manufacturing manager	Inventory planner	Transportation planner / scheduler	Warehouse manager	Customer service manager
Supply chain manager	Production planner	Category / Commodity manager	Logistics engineer	Production manager	Inventory manager	Transportation supervisor	Distribution center manager	
Supply chain consultant	Sales order planner	Purchasing / Procurement manager	Logistics manager			Transportation manager	Business development manager	
Supply chain director	Master scheduler	Purchasing / Procurement consultant	Logistics specialist		Transportation planning / Scheduling manager			
VP supply chain	Demand planner	Purchasing director / procurement	Logistics director					
	Production planner	Director VP purchasing / procurement	VP logistics					
	Demand planning manager							
	Production planning manager							
	Forecasting / Production planning director							

For those professionals already in the field who are looking to improve themselves, certification programs are available, such as the following:

- **Certified Professional in Supply Management (CPSM):** Offered by Institute for Supply Management (ISM) and is recognized internationally
- **Certified Production and Inventory Management (CPIM):** Certification offered by American Production and Inventory Control Society (APICS), and is well known to thousands of companies worldwide
- **Certified Supply Chain Professional (CSCP):** Certification also offered by APICS, and is the most widely recognized credential in the field

You should now have a good understanding of the definition and importance of the field of supply chain management. Next, we examine what I refer to as the planning and scheduling processes that pertain to supply chain operations management.

PART II

Planning for the Supply Chain

3

Demand Planning

I t's only been in the past 20 years or so that businesses have truly come to realize the importance of forecasting. If you think about it, forecasting is usually the first step in the planning and scheduling process for most goods and service organizations, and forecasts for demand drive everything in an organization: from longer-term decisions (3+ years out) as to new facilities and products, to medium-term decisions (months to years out) such as production planning and budgeting, and the short-term (months to a year at most), where we need to know what to produce (or purchase) and deploy (see Figure 3.1).

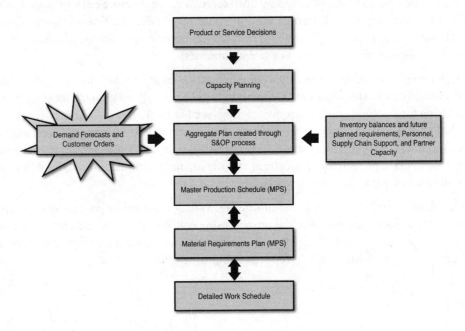

Figure 3.1 Typical planning and scheduling process

The function itself has evolved from being an almost "dreaded" responsibility of sales and marketing, to where operations took control to produce stable production requirements, and on to today where it is most typically part of the supply chain function, where it can rise to the level of importance in an organization to the point where there may be a director of forecasting or demand planning. In fact, there is now a professional organization dedicated to the profession: the Institute of Business Forecasting & Planning (www.ibf.org).

Forecasting Used to Be Strictly Like "Driving Ahead, Looking in the Rearview Mirror"

Historically, manufacturers forecasted sales based on shipments to customers only, which was less than optimal because 1) what we sold may not have been what was ordered (or where it was supposed to ship from) and 2) the true *driver* of most businesses and services is the consumer, not necessarily our distribution channels.

These limitations were due primarily to companies working in more of a "vacuum" because data was hard to get and limited to internal sources and storage space expensive and limited.

Many companies also operated under a *two-number* system, where sales and marketing budgeted one number (which might be changed only once per quarter) and manufacturing developed their own SKU (stock keeping unit) forecast based on more current sales. In some cases, there was even a third number used by those responsible for finished goods deployment to distribution centers, which in many cases was based on percentage allocations of one of the two aforementioned national or global forecasts.

As technology became more readily available (and more affordable) in the mid-1980s to early 1990s, many businesses were able to begin to get a better handle on the forecasting process. Using the *pyramid* approach to forecasting (see Figure 3.2), organizations were able to develop a bottom up/top down *one-number* forecast, which used various statistical methods as well as other sources of information at various levels of detail. These one-number forecasts were able to drive budgeting, production, and deployment simultaneously and be updated typically on a monthly (or more often basis).

The availability and sharing of point-of-sale (POS) data from either paid services or larger customers was also integrated into the process using collaborative programs between manufacturers and retailers such as quick response and efficient consumer response (to be discussed in more detail later) to help reduce the bullwhip effect.

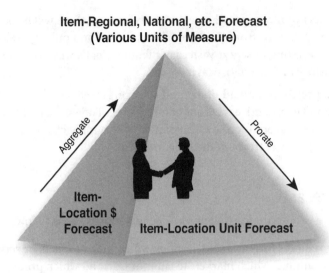

Figure 3.2 The pyramid approach to forecasting

Forecasting Realities

You need to understand certain realities about forecasting before getting into the details of the process:

- **All forecasts are wrong.** It's rare that a forecast is 100% accurate. The idea is to have an integrated, collaborative process that minimizes variance of actual versus target. You'll learn later in this book about the process and importance of setting and measuring forecast accuracy targets.

- **The more "granular" the forecast, the less accurate it is.** A national forecast for a family of items is likely to be more accurate than a weekly forecast for an SKU at a distribution center that handles a region of the country. We can compensate for some of the inaccuracy through proper inventory planning, factoring in scientific safety stock inventory based on desired service levels and reduced lot sizes and cycle times included in Lean, as covered later in the book.

- **It's easier to forecast next month more accurately than next year.** If we know what we sold yesterday, we typically have a better idea of what we'll sell today; whereas 12 months from now, a lot of things can happen that can affect sales.

- **You will get a more accurate forecast using demand history rather than sales history.** Years ago, when data storage costs were high and capacity lower, most companies only stored sales information. Today, most store order or demand information as well. Unless your company has a 100% service level, there will be occasions where you ship short, late, or from the "wrong" location. If you only use sales history, you would be forecasting to repeat yesterday's failure. That's why you should always use demand history to drive statistical forecasts.

- **Forecasting really is a blend of art and science.** As we will discuss, there are both qualitative and quantitative methods of forecasting. Today, the *best practice* is a combination of both, in addition to collaboration with supply chain partners, providing better visibility downstream in the *demand chain*.

Types of Forecasts

Organizations have various forecasting needs. The major ones are as follows:

- Marketing requires forecasts to determine which new products or services to introduce or discontinue, which markets to enter or exit, and which products to promote.
- Salespeople use forecasts to make sales plans, because sales quotas are generally based on estimates of future sales.
- Supply chain managers use forecasts to make production, procurement, and logistical plans.
- Finance and accounting use forecasts to make financial plans (budgeting, capital expenditures, and so on). They also use them to report to Wall Street with regard to their earnings expectations.

Demand Drivers

In general, demand can be driven by a number of internal and external factors, which need to be identified and understood.

Internal Demand Drivers

These types of drivers of demand include sales force incentives, consumer promotions, and discounts to trade. It was only in the past 20 years that some manufacturers and retailers began to better understand the full impact of these drivers on the supply chain resulting in the bullwhip effect.

For example, Procter and Gamble and Walmart have partnered in everyday low pricing (EDLP) to reduce costs and improve service. At one point, P&G went as far as stationing 200 employees at Walmart's headquarters in Bentonville, Arkansas.

Previously, supply chain and operations were at the mercy of these internal drivers and had to live through the consequences. Of course, they will always exist to some degree.

External Demand Drivers

These drivers, although not controllable to any great degree, can be managed better through *best practice* techniques with a structured methodology in place that employs improved communications and integration with other departments within an organization and with customers. These can include events in the environment that are mostly unpredictable, such as terror attacks and stock market crashes, and others that are due to a lack of good communication and visibility, such as new distribution and larger-than-anticipated orders.

Forecasting Process Steps

Everyone does things a little different, so it's always a good idea to develop a standard methodology for a process. In the case of demand forecasting, certain general steps should be included (Heizer & Render, 2013):

1. **Determine the use of the forecast.** Varies by industry and company. In the case of manufacturing, it may be to drive production and deployment, in the case of retail it might be to determine purchasing requirements and for pure service companies might be used primarily for labor staffing.

2. **Select the items to be forecasted.** Will we be forecasting by individual item in various granulations, and at what levels and units of measure will be need to be able to aggregate forecasts and demand history?

3. **Determine the time horizon of the forecast.** Do we need to look at it in the short, medium, or long term (or all of the above), and what type of time planning buckets are appropriate (for example, 30 days or less: daily buckets; 1–3 months out: weekly buckets; and 4+ months: quarterly buckets).

4. **Select the forecasting model(s) and methods.** Based on a number of things we will be discussing, such as where a product is in its *lifecycle*, will we use qualitative, quantitative, or a blend of models? To what degree will we integrate externally supplied information (for example, customer forecasts, POS data, CPFR, and so on), and what weight will we give it?

5. **Gather the data needed to make the forecas.** When using forecasting software, the initial integration will consider much of this, such as using demand versus sales, as mentioned previously, eliminating data errors, and so on. Once this integration has been created and data validated, it becomes more of a maintenance issue for things such as new and discontinued items.

6. **Generate forecasts.** Typically, statistical methods are used to generate a *baseline* forecast, possibly at different levels of detail. The planner will then usually *audit* the results and, if needed, try other statistical models. They will then factor in management overrides based on their experience and knowledge as well as promotional plans, sales estimates, and externally supplied information mentioned earlier.

7. **Validate and implement the results.** During the demand part of the sales and operations planning (S&OP) process that we will discuss in the next chapter, forecasts are reviewed by cross-functional teams at various levels of detail and units of measure to ensure the highest level of accuracy possible. This will ultimately lead to a one-number system that was discussed earlier so that everyone is on the same page.

During this time, recent forecast accuracy will be evaluated, as well, to help target improvement. It is also during this step that the new forecasts are saved to be measured later for accuracy against predetermined variance/error targets, as discussed later in the chapter.

Many companies cycle through this process on a monthly basis (this again varies by industry and how the forecasts will be used), but forecasts are typically adjusted on an as-needed basis due to over/undersells, new demand information, changes to promotions and discounts, and so on.

Quantitative Versus Qualitative Models

There are two general types of forecasting models: quantitative and qualitative.

Qualitative Models

The qualitative method is typically used when the situation is somewhat vague and there is little data that exists. It is useful for creating forecast estimates for new products, services, and technology. Generally, it relies heavily upon intuition and experience.

Qualitative methods include knowledge of products, market surveys, jury of executive opinion, and the Delphi method.

Knowledge and Intuition of the Products

A forecast can come from the experience of a planner/forecaster who has years of experience with the product and can look over historical and forecast statistical estimates to make adjustments based on his or her judgment. This same method can be used with other people in the organization such as sales and marketing to gather their estimates.

However, you must always be aware of biases that may result from different individual's *motivation*. For example, sale personnel may have an *incentive* to hit a high target to reach a bonus.

In my experience as a senior forecaster at Unilever, after awhile I was able to determine that sales estimates were typically 50% high, and once factoring that in, they were fairly useful (at least to start a dialogue). It is also important to be able to share forecast and historical data in units of measure and levels of aggregation that are meaningful to others. For example, the sales department thinks more in terms of revenue dollars, customers, and product categories.

So, a good forecasting process and software system should be able to convert data back and forth to both present the data and receive feedback.

Market Surveys

Market surveys involve the process of gathering information from actual or potential customers. I'm sure most of us have experienced being asked to answer a survey in a mall. When I was an employee at Burger King Corporation at their headquarters in Miami, Florida, we would be asked on occasion to visit the test kitchen upstairs. We would then try different versions of current and new/test items. Usually we were asked to compare items that might have subtle differences, like different brands of ketchup.

Another example are focus groups where people are asked about their perceptions, opinions, beliefs, and attitudes toward a product, service, concept, advertisement, idea, or packaging. Questions are sometimes asked in a group setting, and participants can talk with other group members.

Jury of Executive Opinion

In the jury of executive opinion forecasting method, managers within the organization get together to discuss their opinions on what sales will be in the future. These discussion sessions usually resolve around experienced guesses. The resulting forecast is a blend of informed opinions, with some use of statistical methods.

Delphi Method

In the Delphi method, which is a bit more formal than the jury of executive opinion method, the results of questionnaires are sent to a panel of *experts*. Through an iterative process, multiple rounds of questionnaires are sent out, and the anonymous responses are aggregated and shared with the group at the end of each round. The experts are allowed to modify their answers for each round. The Delphi method seeks to reach the *correct* response through consensus.

Both the Delphi and jury of executive opinion forecasting methods are usually a bit more strategic in nature and used more in developing higher-level longer-terms forecasts.

Quantitative Models

As opposed to qualitative methods, quantitative methods are typically used when the situation is fairly stable and historical data exist. As a result, it is used primarily for existing/current technology products and involves a variety of mathematical techniques we cover in some detail later in this chapter under the two major categories of time series and causal models.

Time Series Models

Time series forecasting uses a set of evenly spaced numeric data that is obtained by observing response variable at regular time periods. The forecasts are based on past values and assume that factors influencing past, present, and future will continue. Relatively simple and inexpensive methods such as moving averages and weighted moving averages are used to predict the future.

Associative Models

Associative (often called *causal*) models forecast based on the assumption that the variable to be forecast (that is, dependent) has a cause-and-effect relationship with one or more other (that is, independent) variables. Projections are then based on these associations. Models such as linear and multiple regression are used in this case.

Product Lifecycles and Forecasting

Before we delve into the various quantitative forecasting models, it is worth discussing the relationship between forecasts and a product's *lifecycle*, because it is somewhat useful to understand where a product is in its lifecycle when determining whether to rely more heavily on qualitative or quantitative models.

Note that the product lifecycle also has an impact on the supply side, as covered in the next chapter.

The phases in the lifecycle of a product or service are introduction, growth, maturity, and decline (see Figure 3.3).

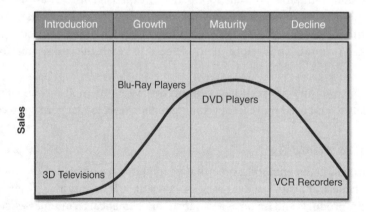

Figure 3.3 Product lifecycle

Introduction

During the introduction phase, there is very little history, if any, to go on, so forecasters tend to rely more on qualitative estimates that are generated both internally and externally. This information can come from sources such as market research; test markets, where that information can be extrapolated; similar items that you've sold before, which may or may not cannibalize other existing items; sales and customer estimates; advance orders to fill the distribution pipeline; and so forth.

Growth

As a product gains momentum through expanded marketing and distribution, some of the simpler time series methods may be used as minimal demand history becomes available.

A general rule of thumb in forecasting is that to generate a decent statistical forecast, you need at least 12 months of history. So, during this growth phase, forecasting is truly a blend of art and science, as both quantitative and qualitative methods are both used to create a blended forecast.

During the growth phase, it can be very easy to over- or underestimate forecasts, which can have dramatic effects on cost and service. So, great care must be taken, and all lines of communication must be established and open both internally and externally, to avoid surprises where possible (which in some cases, such as new distribution, may be hard to avoid).

Maturity

When a product reaches maturity, forecast accuracy tends to improve. For example, when I was in charge of forecasting at Church & Dwight for Arm & Hammer, it was relatively easy to forecast demand for a box of 1-pound baking soda because it had been around for more than 150 years. So, we could rely on simple models to forecast and didn't need as much field information because the item wasn't gaining many new customers. However, once a product reaches maturity, there are opportunities for brand extensions, which is what happened with baking soda. Baking soda actually has hundreds of applications; so, starting with refrigerator and freezer "packs," baking soda gained new life (and new products). This went on to baking soda toothpaste, baking soda deodorant, and so on in the years that followed.

Decline

Once a product goes into its decline phase, besides sales having a general downward trend, the demand locations start to shift because the trend is not uniform. On top of that, other alternative channels not previously used such as dollar stores, discount chains, export, and so on may now be used.

Eventually, the product may be discontinued. However, forecasts must still be generated to run out existing inventory. Therefore, similar to the introductory phase, the forecaster relies more on qualitative than on quantitative methods.

Time Series Components

Time series models can contain some or all of the following components (see Figure 3.4).

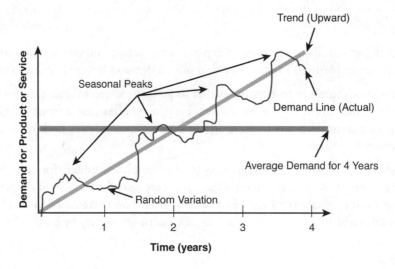

Figure 3.4 Components of demand

- **Trend:** An ongoing overall upward or downward pattern with changes due to population, technology, age, culture, and so on that is usually of several years or more in duration (for example, a fashion trend toward smaller bikinis).

- **Cyclical:** Repeating up and down movements typically affected by business cycle, political, and economic factors, which may vary in length and are usually 2 to 10 years in duration. There are often causal or associative relationships. Examples include economic recessions.

- **Seasonal:** A regular pattern of fluctuations due to factors such as weather, customs, and so on that occur within 1 year. Examples include natural occurrences such as climatic seasons or artificially created such as the school year or seller promotional plans.

- **Random:** Erratic fluctuations that are due to random variation or unplanned events such as union strikes and war, which are usually relatively short in nature and nonrepeatable.

Because these components can be combined in different ways, it is usually assumed that they are multiplicative or additive.

Time Series Models

The most common quantitative time series models are as follows:

- **Naive approach:** Last period's actual demand is used as this period's forecast, without adjusting them or attempting to establish causal factors. (For example, if January sales were 100, then February forecasted sales will be 100.) It is simple, yet cost-effective and efficient.

- **Moving average:** The simple average of a demand over a defined number of time periods and is used if there is little or no trend because it tends to smooth historical data. Typically, more recent history is averaged to create the estimate. (For example, January–March sales are averaged to create an April forecast.)

- **Weighted moving average:** An average that has multiplying factors to give different weights to data at different positions in the sample window. Typically used when some trend might be present because it treats older data as usually less important. The weights are based on experience and intuition and can be used to minimize the smoothing effect if desired.

  For example:

  Weighted moving average forecast for April = .6 * March sales + .3 * February sales + .1 * January sales

  In this example, more weight has been given to March demand than to January and February to generate the April (and onward) forecast.

- **Exponential smoothing:** A smoothing technique used to reduce irregularities. It is a type of the weighted moving average model where weights decline exponentially, with the most recent observations given relatively more weight in forecasting than the older observations. Exponential smoothing requires an alpha smoothing constant (ranges between 0 and 1 and denoted by the symbol α), which is subjectively chosen.

  For example:

  New forecast = Last period's forecast + .7 * (Last period's actual demand – last period's forecast)

  In this example, the smoothing constant used of .7 will give a relatively high weighting or *smoothing* factor to an over- or undersell during the most recent month of history when generating the new forecast.

Associative Models

There are more sophisticated models known as *associative* models, such as linear regression (also known as *least squares method*) and multiple regression analysis, which use the relationship of an independent variable(s) (*x*) to predict a dependent variable (*y*). The reason it is called the least squares method is that the formula draws a *best fit* line through the historical data over time (that is, with the least deviation; see Figure 3.5). That formula can then be used to predict future values of *y*.

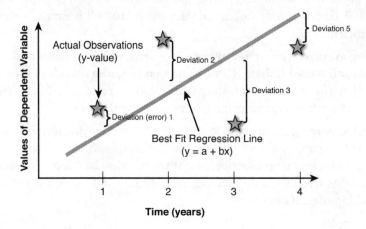

Figure 3.5 Least squares method

In linear regression, the relationship is defined as: *y = a + bx*, where *a* = the *y* axis intercept and *b* = the slope of the regression line.

A simple example of linear regression would be to derive and use this equation to predict future sales (*y*) by plugging in the sales budget (*x*) that we plan on using (with *n* being the total number of observations). If we know that the two variables are strongly correlated, we can easily derive the equation using historical sales personnel employment numbers along with historical sales.

To come up with this equation, we must solve for *b* and then *a*. The formula used to derive each are shown here:

$$b = \frac{\sum xy - n\bar{x}\bar{y}}{\sum x^2 - n\bar{x}^2}$$

$$a = y - bx$$

Once we have solved for *a* and *b*, we have our regression formula and can plug in future sales personnel employment estimates to predict future sales.

Correlation

To measure correlation (that is, the mutual relation of two or more things), we calculate a correlation coefficient, also known as r, which is a measure of the strength and direction of the linear relationship between two variables that is defined as the (sample) covariance of the variables divided by the product of their (sample) standard deviations.

$$r = \frac{n(\sum xy) - (\sum x)(\sum y)}{\sqrt{[n\sum x^2 - (\sum x)^2][n\sum y^2 - (\sum y)^2]}}$$

The range of correlation is 0 to 1. A perfect correlation between two variables would result in an r of +/–1. The lower the correlation between the two variables, the closer to 0 is the result.

Seasonality

In all the previously mentioned time series methods, as well as linear regression, we can apply what is known as a *seasonality index*. As mentioned, this may reflect actual seasonal sales of an item (that is, we sell more snow shovels in the winter) or can be artificially created (for example, a promotional calendar).

A seasonality index is relatively easy to create and can be applied to any of the previously discussed forecasting methods to give the forecast more realistic peaks and valleys.

To create a seasonality index, you must do the following:

1. Calculate an average for all item history (for all years and periods).
2. Average each period's historical data.
3. Divide each period's average by the overall average.
4. Apply the period index to the existing time series or linear regression forecast.

Suppose, for example, that a snow shovel that we sell has historical quarterly sales, as shown here.

Year	Q1	Q2	Q3	Q4	
2010	500	100	25	350	
2011	550	125	15	400	Overall average = 268
2012	625	95	30	325	
2013	550	130	20	450	
Period Average	556	113	23	381	
Index	2.07	0.42	0.09	1.42	

We can create a seasonality index to apply to a *flat* moving average quarterly forecast of 100, for example. To do this, we first calculate an average for each period, and then calculate an overall average of 268. From there, we can calculate indices for each quarter by dividing their period averages by the overall average of 268.

If we had a quarterly forecast for next year of 100/quarter, we could apply the seasonality index for each quarter to that forecast. The resulting quarterly forecasts would be: Q1 = 2.07 * 100, or 207 shovels; Q2 = .42 * 100, or 42 shovels; Q3 = .09 * 100, or 9 shovels; and Q4 = 1.42 * 100, or 142 shovels.

As mentioned previously, seasonality can be natural or induced. In either case, it can change over time and needs to be recalculated on an ongoing basis.

Multiple Regression

When more than one independent variable is going to be used to develop a forecast, linear regression can be extended to multiple regression, which allows for several independent variables. (For example, discounting, promotions, advertising, and so on may all have an impact on sales to one degree or another.) The formula for this is: $y = a + b_1 x_1 + b_2 x_2$ ... (similar to the least squares formula, except with multiple independent variables). This is quite complex and generally done with the help of statistical software.

In the end, you will want to arrive at the best combination of independent variables for the best possible forecast. A statistic called an *r-squared* or *coefficient of determination*, which is the square of the correlation coefficient mentioned earlier and is a measure of the strength of the correlation between y and the various combination of x's, is calculated. The closer to 1.0 the r-squared is, the better the correlation, and hopefully, the more accurate the forecast.

There are many other statistical methods used, ranging from simple to very complex. The best-in-class methods of forecasting use a blend of qualitative and quantitative methods that include collaboration both internally with staff from various departments, including sales, marketing, and finance, and externally with customers and suppliers.

Forecasting Metrics

You cannot control and improve a process if you don't measure it, so it is important to both establish targets and to then track and measure forecast accuracy. There are many ways to establish forecast targets, including historical data, contribution, and so on. The one I prefer is the *ABC method*, which is a way to classify items based on their sales velocity or contribution to profits and can be used to not only set forecasting targets but also in inventory planning and control, as discussed in the next chapter.

To under the ABC method, you needs to understand a phenomenon known as the *Pareto principle* or the *80/20 rule*. It states that a relatively small number of your items generate a fairly large percentage of your sales or profits and are referred to as *A* items (for example, Whopper, fries, and Coke are *A* items at Burger King).

In forecasting, these *A* items require more time and effort put into them and typically have better accuracy as a result. The slower movers, known as *B* and *C* items, are somewhat less important and require less forecasting time and effort and typically have more variability (Myerson, 2014).

Forecast Error Measurement

Forecast accuracy is the difference between what was forecasted for a period and what was actually sold or shipped. It can be measured in whole units or as a percentage.

A number of methods are used to measure and monitor accuracy. The main ones are described in the following subsections.

Mean Absolute Deviation

Simply put, the mean absolute deviation (MAD) is a way to measure the overall forecast error in units over periods of time.

The actual calculation for the MAD is the sum of the absolute error in units divided by the number of occurrences and represented as follows:

$$\frac{(\sum |Actual - Forecast|)}{n}$$

Mean Squared Error

The mean squared error (MSE) is the average of the squared differences between the forecasted and actual values. Its formula is the sum of the square forecast errors divided by the number of occurrences and represented as follows:

$$\frac{\sum (Forecast\ Errors)^2}{n}$$

Mean Absolute Percent Error

As opposed to the MAD and MSE, which can vary in size based on the volume sold of an item, the mean absolute percent error (MAPE) calculates the absolute percentage error. From my experience, this is the most common way for businesses to measure and control forecast accuracy.

Typically, targets are set by ABC code or other methods that highlight relative importance of items, with *A* type items generally having a smaller variance because they are major, everyday items and so more predictable. *C* items, because there are more of them with much smaller volume, tend to be more volatile and so tend to have greater forecast variance.

The MAPE is calculated as follows:

$$\frac{\sum((100*|Actual - Forecast|)/Actual)}{n}$$

Tracking Signal

Over time, forecasts can tend to get out of control fast. As a result, it is a good idea to utilize what is known as a *tracking signal.*

The tracking signal is used to determine the larger deviation (in both plus and minus) of error in forecast, and is calculated by the following formula:

$$\frac{\text{Accumulated Forecast Errors}}{\textbf{Mean Absolute Deviation}}$$

Usually, upper and lower control limits (UCL and LCL) for the number of MADs that the tracking signal represents. There are no "magic" numbers for the UCLs and LCLs, because they are somewhat subjective, but keep in mind that 1 MAD = .8 standard deviations.

In a normal distribution, 3 standard deviations (or +/−4 MADs), should include 99.9% of the occurrences. So, if your tracking signal starts exceeding those levels, it is a good indication that something isn't right.

Demand Forecasting Technology and Best Practices

Computerized forecasting software has been around for a long time. It has evolved from the mainframe to the PC to the Web, from installed applications to cloud based on-demand software-as-a-service (SAAS). These systems range from simple spreadsheet calculations to sophisticated packaged software systems utilizing a variety of forecasting methods and are in some cases integrated with customers and suppliers for improved visibility and collaboration.

Historically, at least in terms of best-of-breed forecasting functionality, the software has been licensed from a separate vendor and integrated with the accounting or enterprise resource planning (ERP) software system. These integrated applications were then used to manage business and automate back-office functions for all facets of an operation, including product planning, development, finance, human resources, manufacturing processes, sales, and marketing. More recently, through development and acquisition, accounting and ERP vendors are increasingly adding this and other nontraditional functionality to their systems (see Figure 3.6).

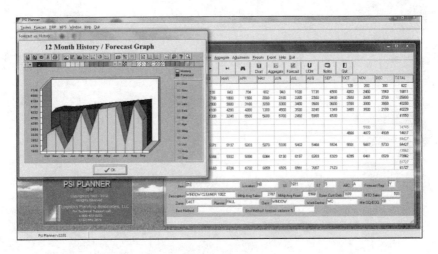

Figure 3.6 Forecasting screen example (PSI Planner for Windows)

Yet, with all of this, according to a KPMG advisory global survey of 544 senior executives (KPMG, 2007), nearly all organizations still use spreadsheets for some parts of the process; more worryingly, 40 percent of them rely solely on spreadsheets to produce the forecast. As a result, this leaves major chances for losses in efficiency and for redundancies in work processes.

However, the KPMG advisory found that the following separated the best in class from the rest regarding forecasting:

- Tend to take forecasting more seriously as they hold managers accountable for agreed-upon forecasts, incentivize managers for forecast accuracy, and use the forecast for ongoing performance management

- Look to enhance quality beyond the basics by incorporating scenario planning and use external market reports and data more often

- Work harder at it by updating and reviewing forecasts more often and more formally and tend to use packaged forecasting software systems more often rather than just spreadsheets

Now that we have answered the demand question, the next step in most goods-oriented planning processes is this: *How much* and *when* do we need to produce or purchase product?

4

Inventory Planning and Control

In most goods and many service organizations such as restaurants, after arriving at a short- to medium-term forecast, we need to figure out how much we want to produce or purchase and when to order or produce it (see Figure 4.1).

This decision is accomplished by determining the current inventory position, which measures a stock keeping unit's (SKU) ability to satisfy future demand. The current inventory position includes scheduled receipts, which are production (or purchase) orders that have been placed but have not yet been received, plus on-hand inventory minus any open customer orders. The current inventory position is then netted against the demand forecast and lead time while adding in buffer or safety stock to create future period (that is, day, week, month, and so on) inventory requirements commonly known as *planned orders*.

These unconstrained requirements are solidified through a process known as *aggregate planning*, covered in the next chapter, which considers various material, manpower, and machine constraints.

Independent Versus Dependent Demand Inventory

There are two general categories of inventory: *independent* and *dependent* demand inventory. This chapter covers independent demand inventory only.

Dependent demand is represented by an item whose demand is linked directly to the demand or production level of another item. Dependent demand items, and the systems for managing them, are typically used in manufacturing. An example of dependent demand inventory requirements would be tires that go on a bike; the production of one bicycle would require two tires of a specific size, and thus the demand for the tires is dependent on the number of bicycles being produced. Material requirement planning (MRP) systems, covered in Chapter 5, "Aggregate Planning and Scheduling," are planning mechanisms to determine requirements for dependent demand.

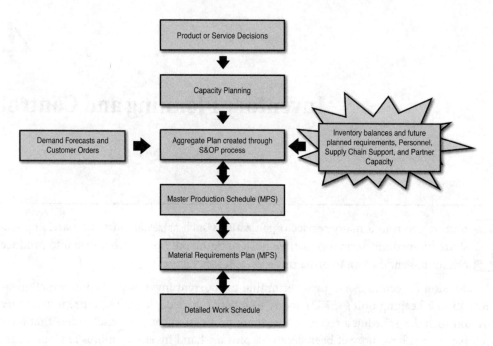

Figure 4.1 Typical planning and scheduling process

By independent demand, however, we are referring to inventory requirements for *finished goods*, which is product ready for the consumer to purchase and use. Finished goods not only exist in a retail store displayed as individual items, but start their journey upstream in the supply chain after production and are typically packed in groups in corrugated containers (for example, 12 bottles/container) because that is more economical for warehouse storage and shipping purposes than storing and shipping one unit of an item.

Finished goods travel through a company's distribution channel prior to getting in the consumer's hands, which typically consists of manufacturers, distributors, or wholesalers and retailers (see Figure 4.2).

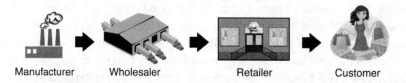

| Manufacturer | Wholesaler | Retailer | Customer |

Figure 4.2 Channels of distribution

A manufacturer produces in lots or batches to gain economies of scale (that is, producing in larger quantities to spread fixed costs over many units so that the cost/unit is lower). A retailer, unless it is large enough to buy in bulk direct from the manufacturers such as Target or Walmart, buys from wholesalers or distributors in smaller quantities.

As product makes its way through the supply chain, both value and cost are added to product.

The value-adding feature of the supply chain was discussed in Chapter 1, "Introduction," with the value chain model and in Chapter 2, "Understanding the Supply Chain," when describing value as a utility.

We talk later in the book about *Lean* concepts for reducing non-value-added activities, which have a major impact on cost and efficiency as well. Suffice it to say, all forms of inventory have cost components, which we will cover shortly.

First, it is important to define the main types of inventory.

Types of Inventory

There are four major types of inventory, as follows:

- **Raw materials and components:** Inventory is usually classified as raw materials if the organization has purchased them from an outside company, or if they are used to make components. This category also includes goods used in the manufacturing process, such as components used to assemble a finished product.
- **Work in process (WIP):** These are materials and parts that have been partially transformed from raw materials but are not yet finished goods and can include partially assembled items that are waiting to be completed.
- **Finished goods:** Products that are ready to be shipped directly to customers, including wholesalers and retailers.
- **Maintenance, repair, and operations (MRO):** These are items a business needs to operate, such as office equipment, packing boxes, and tools and parts to repair equipment.

Costs of Inventory

As inventory works its way from raw material to finished goods, value is added as well as cost. Inventory, as an asset, not only shows up on a financial balance sheet but also goes straight the bottom line on income statements through components of what is known as *holding* or *carrying costs*.

Carrying or Holding Costs

Whether inventory is purchased or produced, costs are involved in the acquisition and holding of it. The components of holding costs are as follows:

- **Capital or opportunity cost (depending on current interest rates can range from 5% to 25%):** Money either has to be borrowed, in which case interest must be paid, or capital from internal sources, which has an *opportunity cost* associated with it (that is, the money would generate a return by investing in other things such as capital equipment).

- **Physical space occupied by the inventory (3% to 10%):** Includes building rent or depreciation, utility costs, insurance, taxes, and so on.

- **Handling of inventory (4% to 10%):** Includes labor cost such as receiving, warehousing, and security and material handling costs, which include equipment lease or depreciation, power, and operating costs.

- **Pilferage, scrap, deterioration, and obsolescence (2% to 5%):** The longer inventory sits around, the more (usually bad) things that can happen to it.

In total, holding costs can range from 15% to 40%, and as you can see, many of the costs are ongoing operating expenses, which can have a significant impact on a business's profitability.

In addition to holding costs, two other major costs are associated with inventory: ordering and setup costs.

Ordering Costs

When placing an order to purchase additional inventory, both fixed and variable are costs involved.

Fixed costs are incurred no matter what and include the cost for the facility, computer system, and so on.

Variable costs associated with purchase orders include preparing a purchase request, creating the purchase order itself, reviewing inventory levels, receiving and checking items as they are received from the vendor, and the costs to prepare and process payments to the vendor when the invoice is received.

Many businesses tend to ignore these costs, especially the variable ones, but those that do calculate it in the range of $50 to $150+/order.

Setup Costs

If you are a manufacturer versus a wholesaler or retailer, there are costs associated with changing production over, known as *setup*, which includes labor and parts as well as downtime.

Note that a full *changeover* includes more than just the equipment changeover and is thought of in Lean terms as "last good piece to first, next good piece," as discussed later in this book.

As with ordering costs, setup costs involve both fixed and variable costs. The fixed costs of setups include the capital equipment used in changing over the production line used for the old items for the new items.

The variable costs include the employee costs for any consumable material used in the teardown and setup. The longer the setup takes, the greater the variable costs.

Total Cost Minimized

The goal is to minimize total costs. Graphically, that occurs at the intersection of holding costs, which go up as lot size quantities increase and setup costs, which go down as the number of orders/setups decrease (see Figure 4.3). So, in effect, holding and setup costs are inverse, resulting in a tradeoff between the two of them. At the point that those costs intersect is where total costs are minimized and is calculated by the simple Economic Order Quantity (EOQ) inventory model.

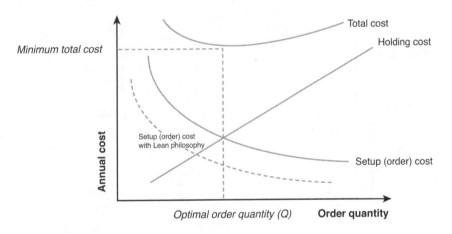

Figure 4.3 Holding versus setup cost tradeoff

As mentioned earlier, it should be noted that there is a lot of pressure to lower inventory costs in an organization. As a result, this pressure ends up falling on the shoulders of the supply chain organization to a great degree. We discuss procurement tactics to minimize some of these costs in Chapter 6, "Procurement in the Supply Chain," but ultimately, process improvement techniques like Lean (Chapter 18, "Lean and Agile Supply Chain and Logistics"), need to be utilized to effectively create a "paradigm shift" of sorts, as shown on Figure 4.3.

Economic Order Quantity Model

The order quantity that minimizes total inventory costs by optimizing the tradeoffs between holding and ordering costs is known as the *economic order quantity* or EOQ. It is one of the most common inventory techniques used to answer the *how much* question.

The EOQ has some assumptions, including the following:

- The ordering cost is constant.
- The rate of demand is known and spread evenly.
- The lead time is known and fixed.
- The purchase price of the item is constant.
- The replenishment is made instantaneously, and the entire order is delivered at one time.

These assumptions can be visualized in terms of inventory usage over time in Figure 4.4, which has become to be known as the *Sawtooth model* (for obvious reasons).

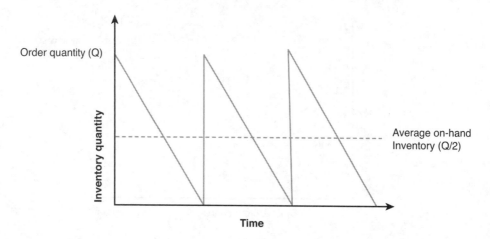

Figure 4.4 Sawtooth model

There are actually three basic EOQ models, the first of which we will mainly discuss and is known as the basic or simple EOQ model.

The other two are as follows:

- **Production Order Quantity model:** As opposed to the basic EOQ model, the Production Quantity model assumes that materials produced are used immediately and as a result lowers holding costs (that is, no instant receipt as in the basic model). As a result, this model takes into account daily production and demand rates.

- **Quantity Discount model:** This is a version of the simple EOQ where pricing discounts are factored into the model based on reaching certain minimum purchase quantities. This then compares the effect of buying more than perhaps is needed but with a lower price, which may offset the impact on holding costs, which in part are based on the price of the product, as you will see in the basic EOQ model.

Basic EOQ Calculation

To calculate the EOQ as well as annual setup, holding, and total inventory costs, we need the following information:

Q = Optimal number of pieces per order (EOQ)

D = Annual demand in units for the inventory item

S = Setup or ordering cost for each order

H = Holding or carrying cost per unit per year

Once we have that information, we can solve for the following:

Annual setup costs = (number of orders placed per year)* (setup or order cost per order) or (D/Q)*S.

Annual holding cost = (average inventory level)* (holding cost per unit per year) or (Q/2)*H.

Total Annual cost = setup cost + holding cost or (D/Q)*S +(Q/2)*H.

Economic Order Quantity (EOQ) = $\sqrt{\dfrac{2DS}{H}}$

Reorder Point (ROP) Models

Now that we've used the EOQ to determine *how much* we need, the next question is *when* to replenish, also referred to as the *reorder point* (ROP).

Basically, two types of models are used in this regard: the Fixed-Quantity (Q) model and the Fixed-Period (P) model.

Fixed-Quantity Model

The Fixed-Quantity, or Q, model has an ROP that is based on inventory reaching a specific quantity (Q), at which point inventory is replenished based on the calculated EOQ (see Figure 4.5).

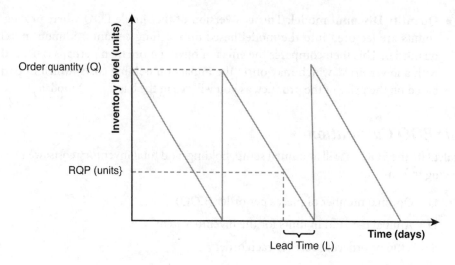

Figure 4.5 Fixed-Quantity (Q) ROP model

The calculation for the ROP = Demand per day × Lead time for a new order (in days) (or $d \times L$).

In a simple example, if our demand is 10 units per day and our replenishment lead time is 3 days, our ROP is 30 units (that is, 10 units × 3 days).

This simplistic model assumes that demand and lead time are constant, which does not reflect reality. So, typically, extra *buffer* inventory is included in this calculation to compensate for this variability, which is known as *safety stock*.

Safety Stock

You can calculate required safety stock in a variety of ways. Many are rules of thumb, and some are statistically based.

In general, the safety stock quantity that is arrived at is *additive* in nature, and so the ROP calculation becomes $d \times L + ss$.

Probabilistic Safety Stock

The idea behind a probabilistic safety stock calculation is that we would like to keep a certain quantity of safety stock to meet a desired service level to compensate for demand variability. If we assume a normal distribution, we can assign a service (or confidence) level as meeting x% of demand during the lead time (see Figure 4.6).

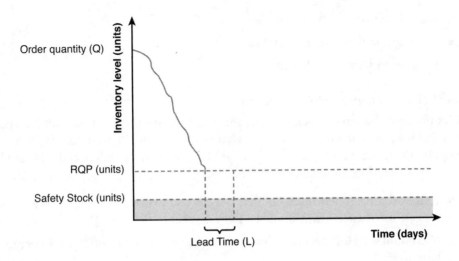

Figure 4.6 ROP with probabilistic safety stock model

To calculate this, we can associate the number of standard deviations around the mean to a confidence level (defined as the number of standard deviations extending from the mean of a normal distribution required to contain x% of the area), which are contained in a commonly available Standard Normal (Z) table. (Some commonly used samples are shown here.)

Z	Confidence Level
1.0	85%
1.3	90%
1.6	95%
3.0	99%

To use this method, we also need to calculate the mean and standard deviation of demand for our item because demand is variable in this case.

Let's take an example where we have a mean demand of 100 units per day, a 1 day lead time, a standard deviation during lead time of 15 units, and a desired service level of 99% (Z = 3.0).

In this type of calculation, the ROP is the expected demand during lead time plus safety stock.

So in our example, the ROP with safety stock calculation would then be 100 + (3.0 × 15) or 145 units.

This model only considers demand variability during lead time only, but there are also other models that compensate for the following:

- Variable demand with constant lead time
- Variable lead time with constant demand
- Variable lead time and demand

Rules of Thumb Safety Stock Calculations

Besides probabilistic safety stock models, some in industry prefer to use rules of thumb instead (which are sometimes referred to as *safety time* because they are expressed in days of supply), which although perhaps not as scientific, are easier to understand and calculate. Some rules of thumb examples include the following:

- **Half lead time:** If demand is 10 units per day and replenishment lead time is 3 days, the calculated safety stock would be 15 units (that is, (10 * 3) / 2).

- **Maximum sales less average sales:** Provides coverage on the *upside* for the occasional large oversell.

- **Statistical safety stock converted to days:** Uses the safety stock probabilistic models' unit calculation above converted to a days of supply inventory target.

Fixed-Period Model

The use of *periods of supply* targets such as in the third example above can be advantageous when you tend to have seasonality with your products, which is one of the main features of Fixed-Period, or P, model (see Figure 4.7).

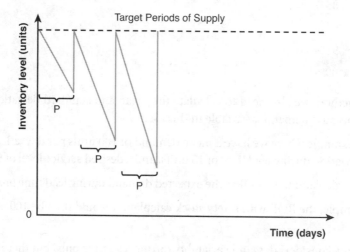

Figure 4.7 Fixed-Period (P) ROP model

In this type of model, inventory is *continuously* monitored. Typically, faster moving items are reviewed more often, with slower movers being reviewed less often.

As opposed to a ROP quantity, individual SKU inventory targets (usually in terms of periods of supply) are the trigger point for replenishment.

Fixed-Period models work well where vendors make routine visits to customers and take orders for their complete line of products, or when it is beneficial to combine orders to save on transportation costs such as shipments to a distribution center. A tool known as *distribution requirements planning* (DRP), which enables the user to set inventory control parameters such as safety stock and calculate the time-phased inventory requirements and is discussed later, is commonly used in the case of managing a network of distribution centers.

Single-Period Model

A Single-Period model is used by companies that order seasonal or one-time items. The product typically has no value after the time it is needed, such as a newspaper or baked goods. There are costs to both ordering too much or too little, and the company's managers must try to get the order right the first time to minimize the chance of loss.

A probabilistic way of looking at this is most helpful. We do this by estimating both the cost of a shortage (Sales price / Unit – Cost / Unit) and of an overage (Cost / Unit – Scrap Value / Unit).

We can then determine a service level (that is, probability of not stocking out) by dividing the cost of shortage by the combined cost of shortage and overage.

The calculated service level percentage can then determine a reorder quantity using the same method as was outlined for the Q ROP model.

ABC Method of Inventory Planning and Control

Many companies treat inventory planning and control with a broad brush, when in fact they should treat items, or least classes of items, differently. A method used in many inventory systems to stratify or classify items is called *ABC analysis*.

ABC analysis is based on the Pareto principle or 80/20 rule, which states that a relatively few number of items typically generate a large percentage of sales or profits (for example, Burger King's Whopper, fries, and Coke versus everything else; see Figure 4.8).

So in terms of inventory planning, the *A* items, because they are the biggest sellers, have a relatively small *days of supply* of inventory target as a result of their high volume and inherent better forecast accuracy as well as the fact that we manufacture or order them more frequently. *C* items typically sell in small amounts and are more volatile, so their inventory target is usually many *days of supply* of inventory, which might not really amount to much anyway because they are small sellers.

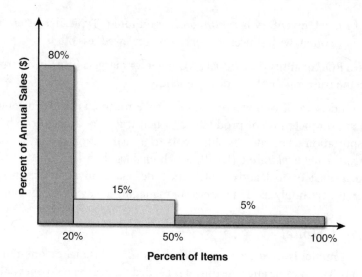

Figure 4.8 Pareto principle or 80/20 rule

From an inventory control aspect, the more important *A* items should have tighter physical inventory control, and the accuracy of inventory records for them should be verified more often.

Realities of ABC Classification

Here are a few of other thoughts regarding the use of ABC analysis for inventory:

- **It's not necessarily a *volume* thing:** It's best to use sales (or cost) dollars or margin versus units when determining ABC codes, because you may sell a high volume of small inexpensive parts and fewer units of much more expensive items. If you used units in that case, the high volume inexpensive items would be given *A* codes, which is the reverse of what you would want to happen.

- **It's not always exactly the 80/20 rule:** It's somewhat subjective as to the cutoffs for the assignments of *A*, *B*, and *C* codes, although it's usually not hard to determine the best cutoff from *A* to *B* and so on.

- **An *A* is not always an *A:*** When you have multiple locations (retail, warehouse, and so on) that stock the same item, it's best to run the ABC analysis by location.

- **History versus forecast for ABC analysis:** History or forecasts can be used to determine ABC codes. Again, it's somewhat subjective, but if the items have history, it's usually best to use that (although do not go too far back because a *B* can become an *A* and vice versa); and if new items, you may be forced to used a forecast, at least for the time being.

Other Uses for ABC Classification

Besides ABC classification's application in forecasting and inventory planning and control, it is also a useful tool for the following:

- **SKU rationalization:** An analysis whereby ABC codes are assigned to determine candidates to be discontinued, scrapped, written off, or sold at a large discount.

- **Quality control:** Pareto charts are one of the tools of quality and used to analyze quality issues where resources are limited and there are a variety of quality issues found that need to be resolved.

Inventory Control and Accuracy

Think of an inventory system in business as being similar to your checking account, where you try to maintain a *perpetual* inventory of your money in a checkbook. Once a month, you get a statement of the *physical* count of your money and then reconcile the two. If there is a discrepancy of any significant size, you have to investigate, find out the reason, and make the adjustment to your checkbook.

Similarly, businesses usually have a software system that keeps a perpetual count of inventory in their factory, warehouse, or store. (Believe it or not, some still do this manually.) Similar to your checking account, system inventory counts can become inaccurate due to inadequate procedures, lost paperwork, and lack of training.

I'm sure that at one time or another you've gone into a store and found that the item you were looking for was out of stock, despite the employee telling you that the system showed that there was plenty in inventory at that location.

It is also critical to manufacturing and sales that inventory counts are accurate in terms of incoming and outgoing recordkeeping and security.

To ensure system accuracy, historically, once per year companies would perform a *physical inventory* count, where everything stops for 2 to 3+ days (that is, nothing comes in and nothing goes out) while employees (usually from another location or temporary workers) go out and physically count the inventory. This is usually done with what is known as a *blind count*, where a person is sent out without knowing the current count, returns his count, and then someone else then reconciles the system perpetual count to the physical count. If it is off, a second person may be sent out to do a *double-blind count*.

Cycle Counting

In recent years, the concept of cycle counting has taken hold. Cycle counts use ABC codes to determine when items should be counted and what the target level of accuracy should be (also referred to as the *ranking method* of cycle counting). Because there are fewer, higher-volume/profit *A* items, they should be *cycled* through more often with extremely high accuracy targets.

Table 4.1 shows an example of a cycle counting schedule.

Table 4.1 Cycle Count Example

Item Class	# Items/Class	Cycle Count Policy	Accuracy Target	# of Items Counted/Day
A	400	Every 20 working days	99%	400 / 20 = 20
B	1,000	Every 60 working days	96%	1,000 / 60 = 17
C	4,000	Every 120 working days	90%	4,000 / 120 = 33

Other benefits of cycle counting include the following:

- Less disruptive to daily operations because it is performed during regular hours with *business as usual.*

- Provides an ongoing measure of inventory accuracy and procedure execution (requiring less safety stock as a result).

- Accuracy issues are corrected on a timelier basis than an annual physical inventory.

- Tailored to focus on items with higher value, higher movement volume, or that are critical to business processes.

- Trained cycle counters perform the work and usually report to an inventory control manager.

- Root cause analysis is used to ensure that once counts are corrected in the system, they don't keep occurring. One method is to determine the cause of the discrepancy and then take counts daily for that item until there are no issues.

Another method of cycle counting is the geographic method. In this method of cycle counting, you start at one end of your facility and count a certain number of products each day until you reach the other end of the building. In this method, you end up counting all your items an equal number of times per year.

Most companies these days do some take on cycle counting, but some still do an annual count, usually at the insistence of the company's auditor for public financial reporting requirements.

Key Metrics

A number of metrics are important to inventory. The most commonly used is inventory turnover. This reflects the velocity at which inventory is flowing through your business and is used both as a budgetary and planning target and benchmark against best-in-class companies for all forms of inventory. The calculation for inventory turnover is Cost of goods sold / Current inventory investment, where *inventory investment* can be represented a number of ways, including the average of several periods (that is, (Beginning plus ending inventory) / 2)), or current on-hand inventory.

For example, if we have $200 million in sales with a cost of goods sold of $100 million and currently have $20 million invested in finished goods inventory, we *turn* our inventory five times per year. This may be good or bad, and that is where benchmarking comes in. If we have a low-cost strategy and the best in class in our industry turn their inventory ten times per year, we have to attempt to turn our inventory faster. We will look at ways to do that when we discuss Lean thinking later in this book.

A high inventory turnover reflects faster-moving inventory, and thus lower holding or carrying costs. The inverse of this, which is commonly used as a target for production and deployment planning, is *periods of supply* (POS), which can be stated in days, weeks, or months of supply. In the earlier example, where our $20 million in finished goods inventory is turned five times per year, we translate that to an average of 2.4 months of supply on hand (that is, 12 months / 5 turns). Again, depending on our POS target, that may be good or bad for our business.

In many cases, a true *depletion* formula is used for POS instead, where current on-hand inventory is run out against future requirements to catch peaks and valleys in demand. (That is, a month of supply for an item may be 100 units in the winter and 1,000 units in the summer.) This is more accurate, but harder to manually calculate.

Many other relevant measures are used, as well, such as *assets committed to inventory* (total inventory investment as a percentage of total assets), *current ratio* (current assets divided by current liabilities), *quick ratio* (current assets less inventory divided by current liabilities), and *gross margin return on inventory* or *GMROI* (gross margin divided by average inventory cost), which is used heavily in retail.

Inventory Planning and Control Technology

Software

As opposed to forecasting software, inventory control software (at least for the perpetual tracking of inventory) is usually included in an accounting or ERP software system as a basic function, although it can be licensed as a standalone system as well.

The basic inventory control systems track the orders, receipts, shrinkage, allocation, and shipment of products. It will produce reports such as current inventory balance, out-of-stock products, and inventory transactions.

Many inventory control systems can also track purchase orders and other inventory value information that is helpful for accounting.

There is a breed of inventory management and control software designed specifically for warehouse operations called *warehouse management system* (WMS) that helps to manage all inventory within the four walls of a warehouse, as discussed in more detail later.

Distribution Requirements Planning Software

A particular type of software called *distribution requirements planning* (DRP) software is more geared to businesses that have to manage a network of distribution centers.

The mechanics of DRP are similar to MRP (materials requirements planning), which is discussed later, in that it develops replenishment plans by evaluating information such as order size, desired safety times/service levels, on-hand inventory, scheduled receipts, and both forecasted and actual demands. However, in the case of DRP, replenishment requirements are for independent demand inventory versus dependent demand for MRP.

DRP compares future demand versus available inventory (plus scheduled receipts such as purchase orders or transfers) to predict future shortages and schedules planned replenishment orders (factoring in lead times) based on user set criteria, including safety stock or safety time targets (see Figure 4.9).

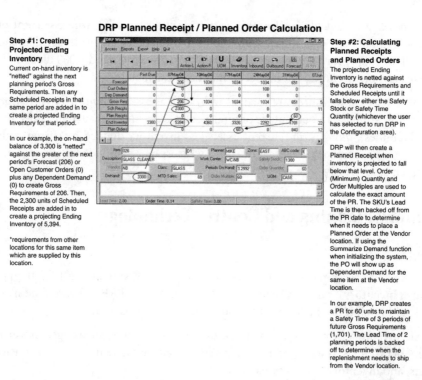

Figure 4.9 DRP screen and description example (PSI Planner for Windows)

DRP is hierarchical because the net requirements can be summarized up the supply chain to the plant level to create the master production schedule (MPS), which can then be *exploded* with a bill of materials (BOM) to generate requirements for raw materials and components.

DRP is ideal for organizations that want to transition from a *push* to a *demand pull* process, resulting in a more-efficient Lean supply chain.

Hardware

Many inventory systems have barcode or radio frequency identification (RFID) functionality to scan items that are received, picked, or transferred. This technology can be used to automate other functions such as cycle counting. The type of equipment required for this includes barcode scanners, radio frequency (RF) tags and readers, mobile handheld computers, and barcode labelers and printers.

Careers

If you are specifically interested in the planning and scheduling topics covered in this and the next chapter, you might want to research information available from various professional organizations.

Both the Council of Supply Chain Management Professionals (www.cscmp.org) and the American Production and Inventory Control Society (www.apics.org) offer certifications in this field, and they have local chapters, both of which can prove useful in terms of education and career advancement.

5

Aggregate Planning and Scheduling

After we've made our best estimate of a demand forecast for goods or services and netted it against our current and targeted inventory position to determine our future inventory requirements, it becomes necessary to make sure that we have enough capacity to meet the anticipated demand.

When we think of planning the capacity for a goods or service business, we typically think in terms of three time horizons:

- **Long range (1–3+ years):** Where we need to add facilities and equipment that have a long lead time.

- **Medium range (roughly 2 to 12 months):** We can add equipment, personnel, and shifts; we can subcontract production and/or we can build or use inventory. This is known as *aggregate planning*.

- **Short range (up to 2–3 months):** Mainly focused on scheduling production and people, as well as allocating machinery, generally referred to as *production planning*. It is hard to adjust capacity in the short run because we are usually constrained by existing capacity.

The supply chain and logistics function must actively support all of these ranges by supplying material and components for production and product to the customer, and in fact, it has many of its own capacity constraints in terms of its distribution and transportation services.

In many service organizations, the actual work of capacity and supply planning for the production of inventory may be partially or totally in another organization, as is the case of retailers or wholesalers. But even in those instances, retail and wholesale supply chain organizations are intertwined with the vendor's manufacturing process. So, they should participate, support, and integrate vendor production plans into their own processes when possible. In addition, service organizations have capacity constraints in terms of various resources that are impacted by inventory levels (labor, warehouse capacity, back room retail storage, shelf space, and so on). Therefore, it is well worth understanding the aggregate planning process no matter where you are in the supply chain.

The Process Decision

Stepping back for the moment, it should be understood that all organizations, both goods and services, have to make what is known as the *process decision*—that is, how the goods or services are to be delivered.

In most established organizations, there is already an existing process that is usually based on the industry's and management's competitive strategy.

Goods and Service Processes

Process choices in goods and service industries can be defined and delineated by what has become to be known as the *product-process matrix* (Hayes & Wheelwright, 1979; see Figure 5.1). In this model, an organization's process choices are based on both the volume produced and variety of products. At the upper left of the chart, companies are considered process oriented or focused, and those in the lower right are considered product focused. The ultimate decision of where a firm locates on the matrix is determined by whether the production system is organized by grouping resources around the process or the product.

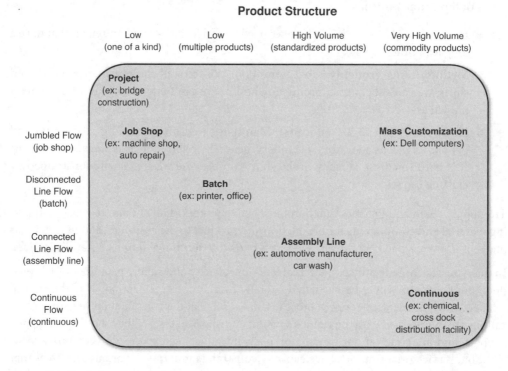

Figure 5.1 Product-process matrix

Project Process

Some industries, such as construction or pharmaceutical, are for the most part project oriented. where they typically make *one-off* types of products. They are usually customer specific and too large to be moved; so people, equipment, and supplies are moved to where they are being constructed or worked on.

Job Shop Process

Job shops typically make low-volume, customer-specific products. Machine shops, tool and die manufacturers, and opticians (that is, prescription glasses) are primary examples of a job shop. As such, they require a relatively high level of skill and experience because they must create products based on the customer's design and specifications.

Each unique job travels from one functional area to another, usually with its own piece of equipment, according to its own unique routing, requiring different operations, different inputs, and requiring varying amounts of time.

Job shops can be extremely difficult to schedule efficiently.

Batch Process

Companies that run a batch process deliver similar items and services on a repeat basis, usually in larger volumes than a job shop. Batch processes have average to moderate volumes, but variety is still too high to justify dedicating many resources to an individual product or service. The flow tends to have no standard sequence of operations throughout the facility. They do tend to have more substantial paths than at a job shop, and some segments of the process may have a linear flow.

Examples of batching processes include scheduling air travel, manufacturing apparel or furniture, producing components that supply an assembly line, processing mortgage loans, and manufacturing heavy equipment.

Assembly Line or Repetitive Process

When product demand is high enough, an assembly line or repetitive process, also referred to as *mass production*, may be used. Assembly line processes tend to be heavily automated, utilizing special-purpose equipment, with workers usually performing the same operations for a production run in a standard flow. In many cases, a conveyor type system links the various pieces of equipment used.

Examples of this include automotive manufacturing (the classic example) and assembly lines. In service industries, examples include car washes, registration in universities, and fast-food operations.

Continuous Flow Process

A continuous flow process, as the name implies, flows continuously rather than being divided into individual steps. Material is passed through successive operations (that is, refining or processing) and eventually come out the end as one or more products. This process is used to produce standardized outputs in large volumes. It usually entails a limited and standardized product range and is often used to manufacture commodities. Very expensive and complex equipment is used, so these facilities tend to produce in large quantities to gain *economies of scale* to spread the considerable fixed costs over as much volume as possible so that the cost per individual pound or unit is as low as possible. Labor requirements are on the low side and typically involve mainly monitoring and maintaining of equipment.

Examples of this include chemical, petroleum, and beverage industries. This type of process is less common in service industries, but a good emerging example in supply chain are cross-dock distribution facilities, which move finished goods product through a distribution facility in as little as 24 to 48 hours.

Mass Customization

Mass customization is a process that produces in high volume and delivers customer-specific product in small batches and can provide a business with a competitive advantage and maximum value to the customer. It is a relatively *new frontier* for most goods and service businesses, and as a result, there aren't that many examples of it.

In manufacturing, Dell computer is a primary example used by many because they allow customers to more or less assemble their own personal computers (PCs) online. Dell then assembles, tests, and ships the PCs directly to the customer in as little as 24 to 48 hours. Some clothing companies manufacture blue jeans to fit an individual customer.

In service industries such as financial planning and fitness, the service is customized specifically to meet the individual needs and therefore is an example of mass customization.

Planning and Scheduling Process Overview

An aggregate plan, also known as a *sales & operations plan* (S&OP), is a statement of a company's production rates, workforce, and inventory levels based on estimates of customer requirements and capacity limitations (see Figure 5.2).

Many service organizations perform aggregate planning in the same way as goods organizations, except that there is more of a focus on labor costs and staffing because it is critical to the service industry (and *pure* service companies don't have inventory to manage, other than supplies).

A variety of methods can be used for aggregate planning, from simple spreadsheets to packaged software using algorithms such as the transportation method of linear programming, which is an optimization tool to minimize costs.

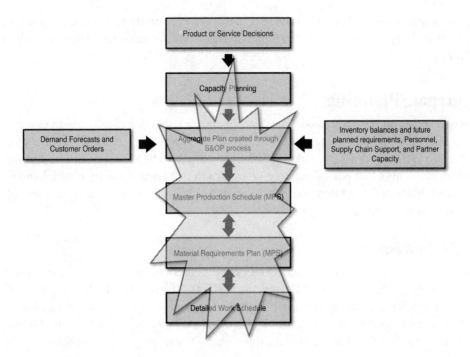

Figure 5.2 Typical planning and scheduling process

Graphical tools can also be used to supplement this process to allow the planner to compare different approaches to meeting demand (see the discussion about *supply options* later in this chapter).

As the name implies, the plan is usually stated in terms of an aggregate, such as product family or class of products, and displayed in monthly or quarterly time periods. It will determine resource capacity to meet demand in the short to medium term (3 to 12 months) and is usually accomplished by adjusting capacity (that is, supply) or managing demand.

Once the aggregate plan is formalized, it is then disaggregated to create a master production schedule (MPS) for independent demand inventory (that is, finished goods), which is also referred to by many as a *production plan*. The MPS is stated in stock keeping unit (SKU) production requirements, usually in daily, weekly, or sometimes monthly time periods.

The MPS is then *exploded* using a bill of materials (BOM), which is basically a *recipe of ingredients* (that is, dependent demand) that goes into the final product (that is, independent demand). This activity is known as *material requirements planning* (MRP).

Once MRP has been run and material availability confirmed, a short-term or detailed work schedule is created. This schedule is where the rubber meets the road because this is a schedule of the actual work to be done, resulting in either meeting or not meeting customer requirements. The work schedule is usually in days or even hours and goes out up to a week or so. It has the specifics as to what product or service will be delivered, when, and who will deliver it.

Aggregate Planning

Aggregate planning, also referred to as sales & operations planning (S&OP), is an operational activity that generates an aggregate plan (that is, for product or service families or classes) for the production process for a period of 2 to 18 months. The idea is to ensure that supply meets demand over that period and to give an idea to management as to material and other resource requirements required and when, while keeping the total cost of operations of the organization to a minimum.

S&OP Process

Best practice companies have a structured S&OP process to ensure success for aggregate/ S&OP planning. The executive S&OP process itself (see Figure 5.3) actually sits on top of the number crunching and analysis being done at a lower level of the organization and involves a series of meetings prior to a final S&OP executive-level meeting, which are used to create, validate, and adjust detail demand and supply plans. The meetings are as follows:

- **Demand planning cross-functional meeting (Step 2):** Generated forecasts are reviewed with a team that may include representatives from supply chain, operations, sales, marketing, and finance. As mentioned in Chapter 3, "Demand Planning," forecasts have been generated statistically and aggregated in a format that everyone can understand and confirm. (For example, sales might want to see forecasts and history by customer in sales dollars.)

- **Supply planning cross-functional meeting (Step 3):** After confirmed forecasts have been *netted* against current on-hand inventory levels to create production/purchasing plans. Again, this data will usually be reviewed in the *aggregate* by product family in units, for example.

- **Pre-S&OP meeting (Step 4):** Data from the first demand and supply meetings are reviewed by department heads to ensure that consensus has been reached.

The discussions from this series of monthly management meetings highlight issues and look at possible resolutions before the outcome of the discussions is presented to the senior management team as a series of issues to be resolved. These issues form the basis of the executive S&OP meeting (Step 5).

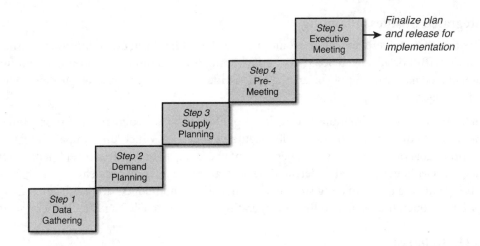

Figure 5.3 S&OP process

The actual aggregate plan requires inputs such as the following:

- Resources and facilities available to the organization.
- Demand forecast with appropriate time horizon and planning buckets.
- Cost of various alternatives and resources. This includes inventory holding cost, ordering cost, and cost of production through various production alternatives such as subcontracting, backordering, and overtime.
- Organizational policies regarding the usage of these alternatives.

Table 5.1 is an example of an aggregate plan for a company that manufactures bicycles.

Table 5.1 Aggregate Plan Example

First Quarter			Second Quarter			Third Quarter		
January	February	March	April	May	June	July	August	September
50,000	30,000	55,000	60,000	80,000	150,000	150,000	125,000	80,000

Some companies start with an aggregate plan and disaggregate to an MPS (that is, SKU level), and others start at the MPS and then aggregate to a class or family of products or services. In any case, the plans, at all levels, including detailed work schedule, are tested for various constraints (manpower, machine, and material) and then adjusted accordingly.

Integrated Business Planning

Note that there is a movement or evolution toward what has been called *integrated business planning* (IBP) or advanced S&OP for some leading organizations, which moves from fundamental demand and supply balancing to a broader, more integrated strategic deployment and management process.

On the operations side, manufacturing develops plans to balance demand and supply but do not always know whether the plan will meet the budgets on which the company's revenue and profit goals are based. The sales department may agree to quotas that meet finance's revenue goals without a detailed understanding of what manufacturing can deliver. IBF attempts to bridge those gaps by making sure that revenue goals and budgets are validated against a bottom-up operating plan, and that the operating plan is reconciled against financial goals.

S&OP in Retail

Also, although S&OP has been a best practice in manufacturing for 25 or so years, the retail industry has been slow to adapt it to their planning processes. The migration toward a broader IBF mentioned earlier for manufacturing may prove to be an impetus to pull retailers into using an S&OP process. In any case, when it is used in retail, the S&OP process is similar to that used by manufacturers. The main differences are that the sponsors and titles of each step as well as the details of each review such as issues, data, and decisions are different.

Demand and Supply Options

During the aggregate planning process, when trying to match supply with demand at the lower cost and highest service, an organization has options to adjust both demand and supply capacity.

Demand Options

These options refer to the ability to adjust customer demand to fit that demand to current available capacity. These options include the following:

- **Influence demand:** This can be accomplished to some degree via advertising, pricing, promotions, and price cuts. Examples including using early-bird meals in a restaurant or discounts offered if you buy before a certain date. These methods might not always have enough of an effect on demand to free up capacity.

  Also, as discussed previously, the use of heavy promotions and discounting can also have the negative *bullwhip effect* as a consequence (thus the reason that some companies have gone to everyday low pricing).

- **Backorders:** These occurs when a goods or service organization gets orders that they cannot fulfill. In many cases, customers are willing to wait. In others, it can result in lost sales. In some industries such as grocery stores, backorders are not used. Instead, if an item is out of stock, it is cut from the order and reordered next time. This is, of course, dangerous if your product is substitutable, because it might not be reordered next time.

- **New or counter seasonal demand:** This can be used to balance demand by season. For example, a company that sells lawn mowers may begin production of snow blowers. Companies must be careful to not go beyond their expertise or base markets.

Supply Capacity Options

These options refer to the ability of an organization to adjust its available resource capacity to meet demand and include the following:

- **Hire and lay off employees:** As demand hits peaks and valleys, flexibility in the workforce can be used to compensate for these fluctuations. Although this can prove beneficial to the company, it can also have risks and costs in terms of unemployment and new-hire training costs.

- **Overtime/idle time:** Most companies have the ability to run some overtime when things get busy. The opposite may be true when things slow down, by moving idle workers to other jobs, at least to some extent. Equipment and workers efforts, to some degree, can also be sped up or slowed down. Although this might extend capacity a bit in the short term, employees may burn out. In the case of slack demand, profitability may suffer as a result of having too many workers doing *make work*.

- **Part-time or temporary workers:** This is especially common for contract manufacturers and in the service industry during the holiday season. It isn't usually an option in more technical jobs, other than some exceptions such computer programming and nursing. Also, quality and productivity may suffer as a result of this approach.

- **Subcontracting (or contract manufacturing):** Very common in some industries, such as cosmetics and household and personal-care products, especially when the demand for a new item is uncertain or a company doesn't yet have the capability to make the product. The downside is that costs may be greater because the subcontractor has to make a profit too, quality may suffer a bit because you have less control, and the fact you may be working with a future competitor.

- **Vary inventory levels:** Inventory may be produced before a peak season when excess capacity may be limited. However, it can also drive up holding costs, including obsolete or damaged inventory. An example of this is the ice cream industry, where ice cream can be produced in the winter and put in a deep freeze until the busy season starts.

Aggregate Planning Strategies

Three general aggregate planning strategies are commonly used, and use many of the demand and supply options discussed earlier:

- **Level plans:** Use a constant workforce and produce similar quantities each time period. This method uses inventories and backorders to absorb demand peaks and valleys and therefore tends to increase inventory holding costs.

- **Chase plans:** This method minimizes finished goods inventories by adjusting production and staffing to keep pace with demand fluctuations. It looks to match demand by varying either workforce level or output rate. This can, of course, negatively affect productivity and costs.

- **Mixed strategies:** Probably used the most with a mix of both of the first two methods. In some cases, inventory is increased ahead of rising demand, and in other cases, backorders are used to level output during extreme peak periods. There may be layoff or furlough of workers during the slower, extended periods, and companies may subcontract production or hire temporary workers to cover short-term peak periods. As an alternative to layoffs, workers may be reassigned to other jobs, such as preventive maintenance, during slow periods.

An example of this is where a company has two production facilities that manufacture the same products, one on the East Coast and one on the West Coast. If one plant has a distinct cost advantage, it may make sense to sometimes shift production to the lower-cost plant and expand its service area temporarily, such as during a slow period of demand. This will, of course, result in less production required at the lower-cost plant during those periods, possibly requiring layoffs. This decision is not to be taken lightly and must consider the total *landed* cost of the product for each plant, including transportation and distribution to the customer.

Master Production Schedule

Once the S&OP process has been completed, the aggregate plan is *disaggregated* into a master production schedule (MPS), which shows net production requirements for the next 2 to 3 months, usually in weekly or monthly time periods by SKU for independent demand items (see Table 5.1). This is known as *time phased planning*.

The net requirements above and beyond existing known ones, which are referred to as *scheduled receipts*, are called *planned orders* and *planned receipts*, the only difference being that planned orders are planned receipts that have been offset by the item's lead time.

Note that the lead time for manufacturing, which is the time required to manufacture an item, is the estimated sum of order preparation time, queue time, setup time, run time, move time, inspection time, and putaway time. In the case of purchased items, the lead time is usually stated by the vendor and may or may not include inbound transit times.

Production Strategies

Manufacturers usually have one or a combination of the following production strategies:

- **Make-to-stock (MTS):** Production for finished goods is based on a forecast using predetermined inventory targets. Customer orders are then filled from existing stock, and those stocks are replenished through production orders. MTO enables customer orders to be filled immediately from available stock and allows the manufacturer to organize production in ways that minimize costly changeovers and other disruptions.

- **Make-to-order (MTO):** Produced specifically to customer order. Usually standardized (but low volume) or custom items produced to meet the customer's specific needs. MTO environments are slower to fulfill demand than MTS and assemble-to-order environments (described next) because time is required to make the products from scratch. There also is less risk involved with building a product when a firm customer order is in hand.

- **Assemble-to-order (ATO):** Products are assembled from components after the receipt of a customer order. The customer order initiates assembly of the customized product. This strategy can prove useful when there are a large number of end products, based on the selection of options and accessories that can be assembled from common components. (This is one example of the concept of *postponement*.)

- **Engineer-to-order (ETO):** This strategy uses customer specifications that require unique engineering design, significant customization, or new purchased materials. Each customer order results in a unique set of part numbers, bills of material (that is, items required to make the product), and routings (that is, steps to manufacture a product).

For the service industry, the MPS may only be an appointment book or log to make sure that capacity (in this case, skilled labor or professional service) is in balance with anticipated demand.

Depending on the production strategy used, the production requirements in the MPS can be expressed based on a forecast, customer orders, or modules that are required for the manufacture of other items (for example, Table 5.2).

Table 5.2 Disaggregation of Aggregate Plan Example

Months								
	January				February			
Aggregate plan quantity	50,000				30,000			
Weeks	1	2	3	4	5	6	7	8
MPS quantity								
26" boys blue	10,000		10,000		5,000		5,000	
12" boys red		12,500		12,500	8,500		8,500	
12" boys yellow		5,000					3,000	

System Nervousness

Frequent changes to the MPS (or subsequently, the material requirements plan, as discussed shortly) can cause what is known as *system nervousness*, where small changes, usually as a result of updating the MPS plan too often, causes major changes to the requirements plan.

To avoid this, many companies use a *time fence*, whereby the planning horizon is broken into two parts:

- **Demand (or firm) time fence (DTF):** A designated period where the MPS is frozen (that is, not changes to current schedule). The DTF starts with the present period, extending as several weeks into the future. It can only be altered by senior management. Unfortunately all too often from what I've seen, the frozen segment is changed often due to *firefighting* and customer emergencies.

- **Planning time fence (PTF):** A designated period during which the master scheduler is allowed to make changes. The PTF starts after the DTF ends and extends several weeks or more into the future.

Material Requirements Planning

After the MPS has been solidified, it can then be *exploded* through a bill of materials (BOM) file to determine raw material and component (that is, dependent demand) requirements.

The information needed to run a Material Requirements Planning (MRP) model includes the MPS, a BOM, inventory balances, lead times, and scheduled receipts (that is, purchase orders and production work orders). All of these inputs need to be accurate and up to date. Otherwise, it's the old *garbage in, garbage out* situation, resulting in poor execution and ultimately customer dissatisfaction.

All the inputs are fairly straightforward, but it would be helpful at this point to delve a little bit into the BOM.

Bill of Materials

A BOM is like a recipe for a product. (In fact, in the case of food, it is.) A BOM file has a defined structure to it. In this structure, the independent demand item is called the *parent* item (for example, 26-inch boys blue bike) and any dependent demand requirements (for example, two wheels for each bike) are called *child* items, with a quantity (2 / Bike in our example) of each child item needed to make each parent item. This is often referred to as the *product structure* (see Figure 5.4).

The finished good or parent item is referred to as being on *level 0* and the child *level 1*. There can be multiple levels in a BOM, in which case the child item on level 1 of the wheel in the bike example can then be the parent to the child items of the rim, tire, spokes (that is, level 2), and so on.

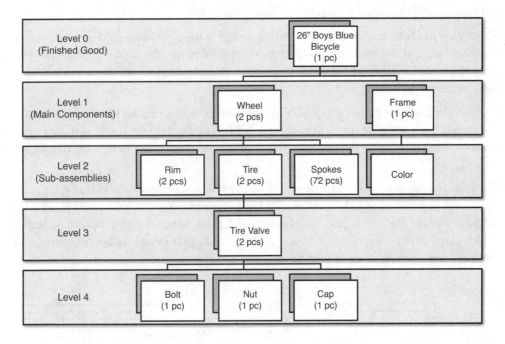

Figure 5.4 Bicycle BOM and product structure

MRP Mechanics

The calculations involved in an MRP system are fairly routine. Think of it as a giant calculator that crunches the information supplied to create net future replenishment requirements based on some user-defined parameters.

As mentioned previously, an MRP system is driven by the MPS (which may, in turn, potentially be driven by a DRP system). The mechanics of the MPS and MRP systems are basically the same, with the requirements from the MPS (independent demand) driving MRP requirements (dependent demand) via the BOM file.

In the bicycle example, Figure 5.5 illustrates the basic calculation where we have gross requirements (in MPS, *gross* is the forecast *consumed* by open customer orders) for the production of 75 bikes in week 8. Typically, safety stock or safety time targets would be in place for independent demand items, but for sake of simplicity, there is none in the example. Because we have 50 bikes in inventory, we need to produce an additional 25 units by week 8. To do so, we need to have 50 wheels and 25 frames available in week 6, after offsetting the components' lead time, for the bike production. Through the BOM explosion, these requirements show up as gross requirements for the wheels and frames in MRP. The same *netting* calculations are then performed to create planned receipts and planned orders for the wheels and frames (and then level 2, level 3, and so on items).

Although it has been said that no safety stock or safety time are required for raw or components because it is factored into finished goods requirements, the reality is that quality and other issues may arise, as well as vendor minimum order quantities, which may call for safety stock as the prudent thing to do.

The actual quantity required is typically rounded up based on various lot-sizing techniques. They range from *lot for lot* (that is, exact requirements no matter how small), which is appropriate for just-in-time (JIT) operations, to economic order quantity (EOQ) calculations, and beyond.

For slow-moving items, an *order time* may be used, which basically states that the planned orders will be grouped together so that one larger order versus many frequent small orders will be placed. In the case of purchased material or parts, vendors may set order minimums (which can always be negotiated). Although this might result in greater holding costs, in the case of slower-moving items, it may be the right thing to do.

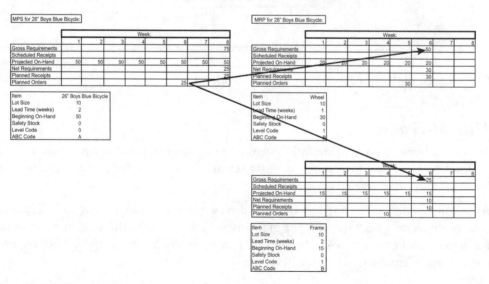

Figure 5.5 MPS and MRP mechanics

Note that in the case of both DRP and MRP, there are *resource* versions (versus *requirement*) that look beyond material requirements and consider other resources impacted such as labor, facilities, and equipment. Some are known as *closed-loop systems*, which allow for the planners to schedule work based on period capacity constraints using smoothing tools that allow the system (manually or automatically) to move requirements around to meet capacity based on priority rules set by the planner such as order splitting (running parts of a work order at two different times) and overlapping (part of a work order can move to a second operation while the rest is still on the first operation).

The planned orders for both independent and dependent demand are then used (either manually or sent electronically to either an enterprise resource planning [ERP] or accounting system) to create production work orders and purchase orders in what is known as *short-term scheduling*.

Short-Term Scheduling

As mentioned before, the short-term schedule (see Figure 5.6) is where the rubber meets the road, because effective schedules are necessary to meet promised customer delivery dates with the highest-quality product or service at the lowest possible cost.

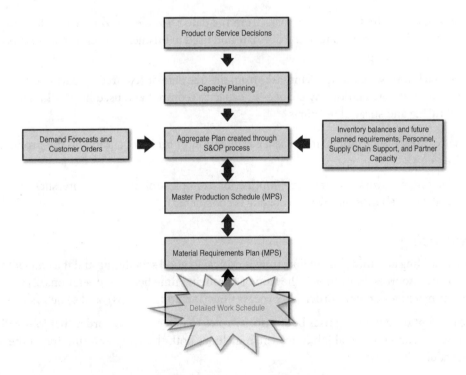

Figure 5.6 Typical planning and scheduling process

Operations scheduling is the allocation of resources in the short term (down to days, hours, and even minutes in some cases) to accomplish specific tasks.

Scheduling includes the following:

- Assigning jobs to work centers/machines
- Job start and completion times
- Allocation of manpower, material, and machine resources
- Sequence of operations
- Feedback and control function to manage operations

Scheduling techniques vary based on the facility layout and production process used.

Effective scheduling can support the supply chain to create a competitive advantage for an organization, as discussed earlier in this book.

Types of Scheduling

Two general types of operations scheduling help to determine the *load* or amount of work that is put through process centers:

- **Forward scheduling:** Plans tasks from the date resources become available to determine the shipping date or the due date and used in businesses such as restaurants and machine shops
- **Backward scheduling:** Plans tasks from the due date or required-by date to determine the start date and/or any changes in capacity required and used heavily in manufacturing and surgical hospitals

In many cases, organizations may use a combination of both depending on the product or service.

The load put on a work center can be *infinite* (for example, unlimited capacity, such as in the basic MRP model) or *finite* (where capacity is considered).

Sequencing

Understanding and minimizing flow time is critical to good scheduling and the efficient utilization of resources. Flow time is the sum of 1) moving time between operations, 2) waiting time for machines or work orders, 3) process time (including setups), and 4) delays.

The concept of sequencing uses both priority rules to determine the order that jobs will be processed in and the actual job time, which includes both the setup and running of the job, to schedule efficiently.

Priority Rules

Although there are many priority rules, including the catchall *emergency* (that is, rush or priority customers), the basic rules are as follows:

- **First come, first served (FCFS):** Jobs run in the order they are received. Perhaps the fairest, although not always most efficient, way of scheduling.
- **Earliest due date (EDD):** Work on the jobs due the soonest.
- **Shortest processing time (SPT):** Shortest jobs run earlier to make sure that they are completed on time. Larger jobs will possibly be late as a result.
- **Longest processing time (LPT):** Start with the jobs that take the longest to get them done on time. This may work well for long jobs, but the others will suffer as a result.
- **Critical ratio (CR):** Jobs are processed according to smallest ratio of time remaining until due date to processing time remaining.

The planner can create schedules based on these methods (manually or automated) to both see the impact on job lateness and flow time and to determine what works best for the company and its customers. It might not always be possible to satisfy all customers, though.

Finite Capacity Scheduling

Finite capacity scheduling (FCS) is a short-term scheduling method that matches resource requirements to a finite supply of available resources to develop a realistic production plan. The MPS and MRP schedules are usually imported into this tool along with other information such as priority rules, setup times, and so on to create short-term daily and hourly schedules.

It uses not only rules-based methods, but also allows for the planner to make up to the minute changes and adjustments as well as perform *what-if* simulation analysis. They allow the planner to handle a variety of situations, including order, labor, and machine changes. The schedules in FCS are usually displayed in Gantt chart form (kind of a sideways bar chart, which can show planned as well as the current status of schedules) and can be accomplished using a range of tools from relatively simple spreadsheets to sophisticated optimization FCS software applications.

Service Scheduling

Although service industries need to schedule production and assembly of product (for example, restaurants), most are primarily interested in scheduling staff. To effectively schedule staffy, they use tools such as appointment systems (to control customer arrivals for service and consider patient scheduling), reservation systems (to estimate demand for service), and workforce scheduling systems (often using seniority and skill sets to manage capacity for service).

These can be manual or automated software systems depending on the size and complexity of the organization.

Technology

Similar to demand planning systems, supply planning tools range from simple spreadsheets (or even the back of an envelope) to sophisticated packaged software systems for optimization.

Much of the basic functionality discussed in this chapter, such as inventory control and management and MRP, is usually part of an organization's ERP or accounting system. Other functions described, such as production and deployment planning and scheduling systems (for example, WMS, DRP, and FCS systems), are not, and may have to be licensed separately as add-ons and integrated with existing ERP and accounting systems.

Now that you have a good handle on the planning and scheduling processes and technologies for the supply chain, it's time to take a look at supply chain management from both a strategic and operational viewpoint.

PART III

Supply Chain Operations

6

Procurement in the Supply Chain

A cquiring materials is the next logical step after the planning process is complete because it is the net result of the planning process just described, whether for raw materials and components for manufacturing or finished goods for wholesalers, distributors, or retailers.

Because purchased materials, components, and services make up a great deal of the supply chain spend for most organizations, resulting in the leverage effect discussed in the first chapter, it is a very visible and important component of supply chain management.

Due to the visibility and general rise in awareness of the importance of supply chain management (SCM) in general, there are many career opportunities in procurement, from the assistant buyer up to director level in many goods and service organizations. The institute for supply management (www.ism.ws) offers the CPSM (Certified Professional in Supply Management), which is helpful both in terms of education and career advancement.

Purchasing is a basic function in most organizations and for the purposes of this book is defined as the transactional function of buying products and services. In a business setting, this commonly involves the placement and processing of a purchase order.

This definition for purchasing is to avoid confusion with two other frequently used concepts and terms of procurement (also known as *sourcing* or *supply management*) and strategic sourcing. We will define them as follows:

- **Procurement:** The process of managing a broad range of processes associated with a firm's need to acquire goods and services in a legal and ethical manner that are required to manufacture a product (direct) or to operate the organization (indirect), the foundation of which is provided by the purchasing function.

- **Strategic sourcing:** The strategic sourcing process takes the procurement process further by focusing more on supply chain impacts of procurement and purchasing decisions, and works cross-functionally within the business firm to help achieve the organization's overall business goals. This includes analysis of the company's annual (or more often) *spend* with suppliers and supply markets and helping to develop a sourcing strategy that both supports the overall business strategy while minimizing cost and risk.

In this chapter, we concentrate primarily on the procurement or supply management process.

Make or Buy

The first decision in this process, at least strategically, is the question of *make or buy*, which is the choice between internal production and external sources.

A simple breakeven analysis can be used to quickly determine the cost implications of a make or buy decision in the following example.

If a firm can purchase equipment for in-house use for $500,000 and produce requested parts for $20 each (assume that there is no excess capacity on their current equipment), *or* they can have a supplier produce and ship the part for $30 each, what is the correct decision: make (assume with new equipment) or buy (that is, outsource production)?

To arrive at the correct decision, a simple breakeven point could easily be calculated as follows:

$$\$500,000 + \$20Q = \$30Q$$
$$\$500,000 = \$30Q - \$20Q$$
$$\$500,000 = \$10Q$$
$$50,000 = Q$$

As the breakeven point is 50,000 units, the answer is that it is better for the firm to buy the part from a supplier if demand is less than 50,000 units, and purchase the necessary equipment to make the part if demand is greater than 50,000 units.

Outsourcing

Many companies choose to outsource activities, resources, and entire business processes for a variety of reasons that include not being viewed as a core competency, high taxes, high energy costs, excessive government regulation, and high production or labor costs. Outsourcing can also sometimes involve transferring employees and assets from one firm to another. Logistics (especially distribution and transportation) is always a good candidate for outsourcing, as are manufacturing and assembly.

The many benefits of outsourcing include the following:

- **Focus on core activities:** Outsourcing noncore activities helps to put the focus back on the core functions of the business, such as sales and marketing.

- **Cost savings:** The lower cost of operation and labor makes it attractive to outsource.

- **Reduce capital expenditures:** Outsourcing frees an organization from investments in technology, infrastructure, and people that make up the bulk of a back-end process's capital expenditure.

- **Increased flexibility:** Outsourcing can improve an organization's reaction to fluctuations in customer demand and changes in technology.

There are also many disadvantages or risks to outsourcing, such as the following:

- **Security risk:** There is always the risk of losing sensitive data and the loss of confidentiality.

- **Loss of management control of business functions:** You may no longer be able to control operations and deliverables of activities that you outsource.

- **Quality problems:** The outsourcing provider may not have proper processes or may be inexperienced in working in an outsourcing relationship.

- **Loss of focus:** The outsourcing provider may work with many other customers, and therefore may not give sufficient time and attention to your company. This might result in delays and inaccuracies in the work output.

- **Hidden costs and legal problems:** This can occur if the outsourcing terms and conditions are not clearly defined.

- **Financial risks:** Bankruptcy and financial loss cannot be controlled if the outsource partner is or becomes financially unstable.

- **Incompatible culture:** Culture of the outsourcing provider and the location where you outsource to may eventually lead to poor communication and lower productivity.

The individual company will have to ultimately make the decision after determining the probability of both the risks and rewards of outsourcing.

Other Supply Chain Strategies

There are a number of other strategies available to the supply chain manager. They can include the following strategies.

In-Sourcing

There is also an opposite, more recent trend in outsourcing and offshoring (the relocation by a company of a business process from one country to another), where companies are starting to perform tasks that were previously outsourced themselves and develop facilities back in their home Western locations, as the results of outsourcing were not exactly as expected (for example, poor quality or low productivity). This is known as *in-sourcing*.

Vertical Integration

A concept similar to in-sourcing but used to develop the ability to produce goods or service previously purchased is known as *vertical integration*. The integration can be forward, toward the customer, or backward, toward suppliers, and can be a strategy to improve cost, quality, and inventory, but requires capital, managerial skills, and adequate demand. It can be risky in industries with rapid changes in technology.

Near Sourcing

There is also a recent trend for U.S. companies called *near sourcing*, primarily as a result of spike in energy costs (that is, transportation) making it more economical to produce closer to home, such as in the Caribbean or Mexico.

Few or Many Suppliers

Companies can choose to go with many suppliers or few suppliers for some materials or products as a supply chain strategy:

- **Many suppliers:** This strategy is used for commodity products in many cases where price is the driving decision factor and suppliers compete with one another.

- **Few suppliers:** In this strategy, the buyer establishes a longer-term relationship with fewer suppliers. The goal is to create value through economies of scale and learning curve improvements. Suppliers are more willing to participate in just-in-time (JIT) programs (a strategy to reduce in-process inventory and associated carrying costs, as discussed later in the Lean chapter of the book) and contribute design and technological expertise. This cost of changing suppliers in this strategy is great because you tend to have all your eggs in one basket and may have invested heavily in this relationship as a result.

Joint Ventures

Joint ventures are formal collaborations between two companies that reduce risk, enhance skills, or reduce costs (or a combination of all three). An example that ended in 2011 was a 50/50 joint venture between Johnson & Johnson and Merck that handled the over-the-counter (OTC) product lines Pepcid, Mylanta, and Mylicon.

Virtual Companies

Virtual companies rely on a variety of supplier relationships to provide services when needed. They usually have very efficient performance, low capital investment, flexibility, and speed. An example of this is Vizio, a company that became the largest-selling brand of LCD television in the United States in 2010, with only 196 employees. They used contract manufacturing and a creative distribution, with the result that a relatively low-cost generic TV could be produced with minimal need for employees or capital.

The Procurement Process

The procurement process typically includes the functions of determining the purchasing specifications, selecting the supplier, negotiating terms and conditions, and issuing and administrating purchase orders.

There are some general steps involved in the procurement process, which we will review in some detail (see Figure 6.1). They are as follows:

1. Identify and review requirements.
2. Establish specifications.
3. Identify and select suppliers.
4. Determine the right price.
5. Issue purchase orders.
6. Follow up to assure correct delivery.
7. Receive and accept the goods.
8. Approve invoices for payment.

Figure 6.1 The procurement process

Identify and Review Requirements

When discussing requirements, you need to understand that procurement activities are often split into two categories (direct and nondirect) depending on the consumption purposes of the acquired goods and services (see Table 6.1).

The first category, direct, is production-related procurement, and the second is indirect, which is non-production-related procurement.

Direct procurement is generally referred to in manufacturing settings only. It encompasses all items that are part of finished products, such as raw material, components, and parts. Direct procurement, which is a major focus in supply chain management, directly affects the production process of manufacturing firms. It also occurs in retail where *direct spend* may refer to what is spent on the merchandise being resold.

In contrast, *indirect procurement* activities concern operating resources that a company purchases to enable its operations (that is, maintenance, repair, and operations [MRO] inventory defined in Chapter 4, "Inventory Planning and Control," as well as capital spent on plant and equipment). It comprises a wide variety of goods and services, from standardized low-value items, such as office supplies and machine lubricants, to complex and costly products and services, such as heavy equipment and consulting services.

		TYPES		
		Direct procurement	Indirect procurement	
		Raw material and production goods	Maintenance, repair, and operating supplies	Capital goods and services
FEATURES	Quantity	Large	Low	Low
	Frequency	High	Relatively high	Low
	Value	Industry specific	Low	High
	Nature	Operational	Tactical	Strategic
	Examples	Resin in plastics industry	Lubricants, spare parts	Resin and plastic product storage facilities

Table 6.1 Direct Versus Indirect Procurement

The source for requirements can come from material requirement planning (MRP) systems via planners and purchase requisitions from other users in the organization. (A purchase or material requisition is a document generated by an organization to notify the purchasing department of items it needs to order, the quantity, and the time frame that will be given in the future.)

During this step, purchasing will review paperwork for proper approvals; check material specifications; verify quantity, unit of measure, delivery date and place; and review all supplemental information.

Establish Specifications

To establish specifications, you must identify quantity, pricing, and functional requirements as described here:

- **Quantity:** In the case of small-volume requirements, you need to find a standard item. If larger volume, it must be designed for economies of scale to both reduce cost and satisfy functional needs.
- **Price:** This relates to the use of the item and the selling price of the finished product.
- **Functional:** There is a fundamental need to understand what the item is expected to do per the users. This includes performance and aesthetic expectations (for example, hand can opener; how smoothly does it remove the top of cans as well as how ergonomically appealing is the design?).

In general, the description of the item may be by brand or specification. You use brand if the quantity is too small or if the item is patented or is requested by a customer. It is by specification if you're looking for very specific physical or chemical makeup, material, or performance specifications.

The source of the specifications themselves can be based on buyer requirements or standards that may be set independently.

If the buyer sets the specifications, it can become a long and expensive process requiring detailed description of parts, finishes, tolerances, and materials used, resulting in the item being expensive to produce.

Standards, in contrast, set by government and nongovernmental agencies, can be much more straightforward to use because they tend to be widely known and accepted, lower in price, and more adaptable to customer needs.

Identify and Select Suppliers

The next step in the procurement process is to identify and select suppliers. Typically, this involves coming up with a *long list* of suppliers who meet your requirements in general, and then whittling the list down to final candidates before selecting the ultimate vendor.

Identification of potential suppliers can come from a variety of sources, including the Internet, catalogs, salespeople, and trade magazine and directories.

After you have identified potential vendors, a *request for information* (RFI) is issued to them that states a bit about your company and its requirements as well as requested background on the vendor. It's usually not too difficult to refine the vendors that respond down to a smaller list of candidates (usually five to ten), and from there it is best to include a multifunctional team of employees to determine the finalists.

Once you have it down to a short list, a *request for quotation* (RFQ) or *request for proposal* (RFP) is issued. An RFQ is an invitation to selected suppliers to bid or quote on delivering specific products or services and will include the specifications of the items/service. The suppliers are requested to return their bids by a set date and time to be considered for selection.

Discussions may be held on the bids, in many cases to clarify technical capabilities or to note errors in a proposal. The initial bid does not have to mean the end of the bidding because there may be more than one round.

Vendor Evaluation

I've found what is known as the *factor rating method* (see Figure 6.2) to be useful in the task of vendor evaluation.

Criteria	Weights	Scores (1-5)	Weight x Score
Engineering and research capabilities	0.10	4.00	0.40
Production process capability (flexibility/agile)	0.15	5.00	0.75
Delivery capability	0.05	3.50	0.18
Quality and performance	0.20	3.00	0.60
Location	0.05	1.00	0.05
Financial and managerial strength (stability and cost structure)	0.15	5.00	0.75
Information systems capability (e-procurement, ERP)	0.10	2.00	0.20
Reputation (sustainability/ethics)	0.10	5.00	0.50
Total	1.00		3.43

Figure 6.2 Factor rating method for vendor evaluation

The factor rating method identifies criteria that need to be considered as part of what you will be buying and assigns weights as to the relative importance of each of these factors. You then score how well each supplier compares on each factor and give them a score, which is weighted times the rating.

Although this might not be the total decision making factor, it can get you close enough to help you make a final decision. *Intangible* factors can always come into play, such as personal opinions of executives, prior experience with a vendor, and so on.

Many factors besides price (and not always the lowest is selected) are important when selecting a supplier, such as the following:

- **Technical ability:** As their product will become part of your product, can they help you to develop and make improvements to your product?
- **Manufacturing capability:** Can they consistently meet your stated quality and specifications?
- **Reliability:** Are they reputable and financially stable?
- **After-sales service:** Do they have a solid service organization that offers technical support?
- **Location:** Are they close enough to support fast and consistent delivery and support service when needed?

Determine the Right Price

As pointed out before, although price might not be the only determinant, it certainly contributes greatly to the bottom line because it can be upward of 50% of the cost of goods sold.

Three basic models are used as a basis for pricing:

- **Cost based:** The supplier makes their financials available to purchaser.
- **Market based:** The price based on published, auction, or indexed price.
- **Competitive bidding:** This is typically used for infrequent purchases, but can make establishing a long-term relationship more difficult.

When preparing to negotiate price, preparation is the key. On a personal level, if you are buying a house or car, the more research you do, the better idea you have of what is available and what is a *fair* price in the market area (at least to you). Thanks to the Internet, many sources are available to get a good idea as to what's available and a range of pricing based on recent history. The same goes for business negotiations, where the buyer should have knowledge of the seller's costs to some extent.

Negotiation

For the most part, negotiations are based on the type of product:

- **Commodities:** The price is usually determined by the market.
- **Standard products:** The price is set by catalog listings, and there is usually little room for negotiation (other than volume).
- **Small value items:** Companies should try to reduce ordering costs or increase volume where possible.
- **Made-to-order items:** Prices are based on quotations from a number of sources, and as a result, prices are negotiated where possible.

Where negotiations are possible, two general types of negotiation can be used: distributive and integrative.

In distributive bargaining, the goals of one party are in fundamental, direct conflict to another party, resources are fixed and limited, and maximizing one's own share of resources is the goal for both parties. So, in this case, there is usually a *winner* and a *loser*.

You need to set a target point and a walk-away point to negotiate a final price that is satisfactory to the buyer. Determining these may take a good amount of research and judgment. The seller may have a listing or asking price, and you will submit an initial offer or counteroffer. This type of negotiating usually requires sufficient clout to justify lower pricing. Larger companies with multiple locations or business units may have sufficient volume to justify this.

When I was a member of General Electric's corporate sourcing, we were able to leverage over $1 billion/year spent annually on transportation corporation-wide by collecting freight volumes by mode for all the 100+ GE units to negotiate significant savings. This was accomplished not only by collecting and analyzing the annual spend, but also by reducing the number of carriers within each mode to a company-wide group of *core* carriers to maximize negotiation power.

Integrative negotiation, in contrast, is more collaborative, with a goal for a *win-win* conclusion by the creation of a free flow of information and an attempt to understand the other negotiator's real needs and objectives. This process emphasizes commonalties between the parties and minimizes the differences through a search for solutions that meet the goals and objectives of both sides.

Issue Purchase Orders

At this point, we move from procurement to more of the day-to-day supplier scheduling and follow-up, which go more under the heading of purchasing activities. This involves execution of the master schedule and MRP to ensure good use of resources, minimize work in process (WIP), and provide the desired level of customer service. This usually falls under the auspices of what is known as a buyer/planner, who works hand in hand with the master scheduler. Buyer/planners are responsible for the control of production activity and the flow of work through the plant and can be also be responsible for purchasing, materials requirements planning, supplier relationship management, product lifecycle and service design, and more. They also coordinate the flow of goods from suppliers.

The purchase order is used to buy materials between a buyer and seller. It specifically defines the price, specifications, and terms and conditions of the product or service and any additional obligations for either party. The purchase order must be delivered by fax, mail, personally, email, or other electronic means.

The types of purchase orders may include the following:

- **Discrete orders:** Used for a single transaction with a supplier, with no assumption that further transactions will occur.
- **Prenegotiated blanket:** A purchase order made with a supplier containing multiple delivery dates over a period of time, usually with predetermined pricing, which often has lower costs as a result of greater volumes (possibly through centralized purchasing and/or the consolidation of suppliers) on a longer-term contract. It is typically used when there is an ongoing need for consumable goods.
- **Prenegotiated vendor-managed inventory (VMI):** The supplier maintains an inventory of items at the customer's plant and the customer pays for the inventory when it is actually consumed. Usually for standard, small-value items such as maintenance, repair, and operating supplies (MRO) like fasteners and electrical parts.

- **Bid and auction (e-procurement):** This involves the use of online catalogs, exchanges, and auctions to speed up purchasing, reduce costs, and integrate the supply chain. There are many e-commerce sites for industrial equipment and MRO inventory auctions; they vary in format from catalog (for example, www.grainger.com, www.chempoint.com) to auction (for example, www.biditup.com). Websites can be for standard items or industry specific.

- **Corporate purchase card (pCard):** This is a company charge card that allows goods and services to be procured without using a traditional purchasing process; sometimes referred to as *procurement cards* or *pCards*. There is always some kind of control for each pCard, such as a single-purchase dollar limit, a monthly limit, and so on. A pCard holder's activity should be reviewed periodically independently.

To further enhance the speed and accuracy of transactions, many companies use what is known as *EDI* (electronic data interchange), which is the computer-to-computer exchange of business documents in a standard electronic format between business partners. In the past, EDI transactions either went directly from business to business (in the case of large companies) or through third parties known as *value-added networks* (VANs). Today, a large portion of EDI transactions now flow through the Internet.

Sometimes included in the category of EDI is the use of electronic funds transfer (EFT), which is the electronic exchange transfer of money from one account to another, within a single financial institution or across multiple institutions, through computer systems. This also includes e-commerce payment systems, which facilitate the acceptance of electronic payment for online transactions, which has become increasingly popular as a result of the widespread use of the Internet-based shopping and banking.

Follow Up to Ensure Correct Delivery

Enterprise resource planning (ERP) software modules assume that scheduled dates will be received on time. However, a scheduled delivery date must be monitored and managed to identify and avoid possible missed dates in advance where possible. In some cases, delays may be inevitable, and as a result, recovery plans must be developed and managed.

To collaboratively resolve problems, it is also critical to understand the supplier's production process, capacity, and constraints.

On occasion, expediting is necessary, but it should be on an exception basis. Supplier performance should be monitored on an ongoing basis. If a particular supplier is consistently being expedited, the corrective action occur.

In many organizations, purchasing may work hand in hand with either their traffic or transportation department or that of the vendors (depends on shipping terms, as discussed later).

Receive and Accept Goods

The key objective at receipt of goods is to ensure that proper physical condition, quantity, documentation, and quality parameters are met. To accomplish this requires a cross-functional activity among purchasing, receiving, quality control, and finance.

Receiving is technically a *non-value-added* activity from a customer perspective because it is designed to ensure that everything up to that point has been done properly. The goal is to ensure quality throughout and to reduce or eliminate the need for inspection. In many cases, technology such as barcode scanners and handheld computers can automate the process. Some of the inspection processes can also be reduced or eliminated by various inspection and certification processes being performed by the vendor.

Approve Invoice for Payment

The final step in the procurement process is approving an invoice for payment (see Figure 6.3) according to the terms and conditions of the purchase order (PO). Typically, the data in the PO is matched with that found on the packing slip that was received and checked when the product arrived and the invoice.

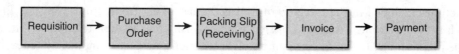

Figure 6.3 Document flow

Any discrepancies must be reconciled before payment is issued to the vendor. In some cases, small levels of discrepancies can be ignored (for example, +/– 3% or +/– $20).

Discounts for early payment should be taken whenever possible; although in a sluggish economy, many customers try to extend payment as long as possible because of cash-flow issues.

Key Metrics

Price is only one measure of cost and only one element of assessing the attractiveness of a supplier, but it is the most common way most companies view and manage interactions with their suppliers.

Besides price (which is benchmarked against industry standards in many cases), most companies also spend a lot of time and attention on operational dimensions of measurement, which can include quality measures such as parts per million defect rates, and service-level measures such as time to respond to inquiries, on-time delivery, and so on.

Some forward-thinking companies are beginning to share information about their strategic business strategies with their key suppliers and have joint discussions about how they can contribute and what metrics can be used to evaluate those contributions. These metrics can be designed to highlight areas for supplier improvement or development, to provide early warning of potential problems at suppliers, and to ensure the ongoing financial health and sustainability of key suppliers.

Technology

Most ERP and accounting systems have at least some purchasing features, at least to create purchase orders directly or from an MRP system.

There are also Internet applications such as e-commerce sites, exchanges, and auctions, as mentioned previously, for e-procurement.

There are also web-based applications (Ariba software, for example) that enable companies to facilitate and improve the procurement process by providing solutions that help companies analyze, understand, and manage their corporate spending to achieve cost savings and business process efficiency. Ariba started off with the idea of purchasing staff buying items from vendors who provided their catalogs online, because the typical procurement process can be labor intensive and often costly for large corporations. Customers are offered a large number of supplier catalogs to purchase from.

Ariba software enables a company to automate, monitor, and control the complete purchasing lifecycle from requisition to payment. Users can create requisitions that are approved according to preconfigured business rules that each company decides upon. Purchase orders can be automatically generated and sent directly to suppliers, and order acknowledgments and ship notices are sent back to the original requestor.

The invoicing process is relatively easy for the suppliers using a tool such as Ariba because they can create an invoice directly from the requestors purchase order. Invoices are then prematched with the purchase order line items and any receiving information so that the requestor can reconcile and pay without any delays.

We cover transportation next, and then distribution operations, both of which are necessary to keep the entire supply chain process, as discussed so far, moving.

7

Transportation Systems

s the saying goes, "A chain is only as strong as its weakest link." In the case of the supply chain, the link is your transportation system, and its strength can mean the difference between the success and failure of your business.

To be successful, the transportation system used to connect your supply chain must be managed and controlled properly, with complete visibility and great communication between partners. Transportation and logistics costs (mainly warehouse operations, which are covered in the next chapter) can account for as much as 7% to 14% of sales depending on the industry you are in. Transportation costs alone comprise the vast majority of this expense for most companies. Best-in-class companies have transportation and logistics related costs in a range of 4% to 7% depending on industry sector. So, it's not hard to see both operationally and financially important transportation is to a successful business.

There are also many professional career opportunities in the transportation field, both in corporate goods and services organizations (usually in the transportation, traffic, or logistics departments) as well as in transportation companies. They can range from corporate to operations and even sales. Besides the Council of Supply Chain Management Professionals (CSCMP) organization mentioned previously, there are many others, such as AST&L (American Society for Transportation & Logistics; www.astl.org) and SOLE (International Society for Logistics; www.sole.org).

As a first step, it is important to understand background on the history of transportation systems in the United States, followed by a discussion of the various characteristics of transportation types and modes, along with their cost elements, rate structures, and some of the necessary documentation.

Brief History of Transportation Systems in America

In the late 18th century, overland transportation was primarily by horse, and water and river transportation was primarily by sailing vessel.

As a result of the distances between cities and the cost to maintain roads, many highways in the late 18th century and early 19th century were actually privately maintained turnpikes. Other highways were largely dirt roads and impassable by wagon for at least some of the year. Economic expansion in the late 18th century to early 19th century was the impetus for the building of canals to move goods rapidly to market. One of the most successful examples was the Erie Canal.

As a result, access to water transportation tended to shape the geography of early settlements and boundaries.

Development of the midwestern and southern states located on or near Mississippi River system was accelerated by the introduction of steamboats on these rivers in the early 19th century.

The rapid expansion railroad transportation in the 1830s to 1860s ended the canal boom and provided a timely, scheduled year-round mode of transportation. Railroads rapidly spread to connect states by the mid-1800s. As the United States industrialized after the Civil War, and with the creation of the transcontinental rail system in the 1860s, railroads expanded rapidly across the United States to serve industries and the growing cities.

The passage of the Act to Regulate Interstate Commerce, in 1887, allowed the federal government to become more active in protecting the public interest. It established the Interstate Commerce Commission (ICC), which had broad regulatory power in the area of transportation, which lasted until 1980. During this period, the federal and state governments determined who could provide transportation services and the price they could charge for their services.

The invention of the automobile started the decline of passenger railroads and increased mobility in the United States, the latter adding to economic output.

Freight railroads also began to decline as motor freight captured a significant portion of the business. This loss of business, along with the highly regulated environment with its restricted pricing power, forced many railroads into bankruptcy and resulted in the nationalization of several large eastern carriers into the Consolidated Rail Corporation (Conrail).

Air cargo deregulation was signed into law a year prior to the passage of the Air Deregulation Act in 1978, which was directed at passenger airlines. Deregulation in trucking and railroads became official with the passage of the Motor Carrier Act of 1980 (MC 80) and the Staggers Rail Act (Staggers Act), which created a regulatory environment favorable to the business economics of the railroad and trucking industries.

With globalization starting in the 1980s, air cargo transport, a vital component of many international logistics networks commonly used for perishables and premium express shipments, grew rapidly.

In the 1990s, the increase in foreign trade and intermodal ocean container shipping led to a resurgence of freight railroads, which today have consolidated into two eastern and two western private transportation networks: Union Pacific and BNSF in the west, and CSX and Norfolk Southern in the east. The Canadian National Railway acquired the Illinois Central route down the Mississippi River Valley.

Transportation Cost Structure and Modes

The primary modes of transportation are truck/motor carrier, rail, air, water, intermodal transportation, and pipeline. Before getting into specifics, it is helpful to understand the cost structure for transportation because a primary source of difference between modes is operating cost and flexibility.

Transportation Costs

Transportation costs are both fixed and variable. The fixed-cost component refers to costs that do not change with the volume moved, such as buildings, equipment, and land. Variable costs, in contrast, are costs that do change with the volume moved, such as fuel, maintenance, and wages.

The areas where these costs occur in transportation are as follows: 1) the *ways* (that is, road, air, and water), 2) terminals (including administration) where goods are loaded and unloaded, and 3) the vehicles themselves used to haul the freight.

Ways

The *ways* are the land, water, road, space, and so on over which goods are moved and may be owned by the operator (railroad tracks), run by the government (roads, canals), or made by Mother Nature (ocean).

Terminals

The terminals are used to sort, load and unload goods, connect between line-haul and local deliveries or between different modes or carriers as well as dispatching (that is, to monitor the delivery of freight over long distances and coordinate delivery pickup and drop-off schedules), maintenance, and administration.

Vehicles

The vehicles themselves are either owned or leased by the transportation companies and have a mix of fixed (for example, vehicle capital investment) and variable (for example, fuel, maintenance, and labor) operating costs.

Modes

In general terms, trucks carry the greatest dollar volume of freight in the United States (71%) because they tend to haul higher-value consumer goods. Rail, which tends to haul lower-value commodity items longer distances, matches motor carriers in terms of ton-miles (39% each). Not surprisingly, air transport has by far the longest average miles per shipment, at 1,304, followed by rail (728) and then water (520) (U.S. Department of Transportation, 2007; see Figure 7.1).

Table 1070. Shipment Characteristics by Mode of Transportation: 2002 and 2007

[8,397,210 represents $8,397,210,000,000 (except as indicated otherwise). For business establishments in mining, manufacturing, wholesale trade, and selected retail industries. 2007 industries classified by the 2002 North American Industry Classification (NAICS). 2002 industries classified by the 1997 North American Industry Classification. Selected auxiliary establishments are also included. Based on the 2007 Economic Census; see Appendix III]

Mode of transportation	Value (mil. dol.)		Tons (1,000)		Ton-miles (mil.)		Average miles per shipment	
	2002	2007	2002	2007	2002	2007	2002	2007
All modes	**8,397,210**	**11,684,572**	**11,667,919**	**12,543,425**	**3,137,998**	**3,344,658**	**546**	**619**
Single modes	**7,049,383**	**9,539,037**	**11,086,660**	**11,696,128**	**2,867,938**	**2,894,251**	**240**	**234**
Truck [1]	6,235,001	8,335,789	7,842,836	8,778,713	1,255,908	1,342,104	173	206
For-hire truck	3,757,114	4,955,700	3,657,333	4,075,136	959,610	1,055,646	523	599
Private truck	2,445,288	3,380,090	4,149,658	4,703,576	291,114	286,457	64	57
Rail	310,884	436,420	1,873,884	1,861,307	1,261,612	1,344,040	807	728
Water.....................	89,344	114,905	681,227	403,639	282,659	157,314	568	520
Shallow draft	57,467	91,004	458,577	343,307	211,501	117,473	450	144
Great lakes	843	(S)	38,041	17,792	13,808	6,887	339	657
Deep draft	31,034	23,058	184,610	42,540	57,350	32,954	664	923
Air (includes truck and air).......	264,959	252,276	3,760	3,611	5,835	4,510	1,919	1,304
Pipeline [2]	149,195	399,646	684,953	650,859	(S)	(S)	(S)	(S)
Multiple modes	**1,079,185**	**1,866,723**	**216,686**	**573,729**	**225,715**	**416,842**	**895**	**975**
Parcel, U.S. Postal Service or courier.....................	987,746	1,561,874	25,513	33,900	19,004	27,961	894	975
Truck and rail............	69,929	187,248	42,984	225,589	45,525	196,772	1,413	1,007
Truck and water..........	14,359	58,389	23,299	145,521	32,413	98,396	1,950	1,429
Rail and water	3,329	13,892	105,107	54,878	114,985	47,111	957	1,928
Other multiple modes	3,822	45,320	19,782	113,841	13,788	46,402	(S)	1,182
Other and unknown modes ...	**268,642**	**279,113**	**364,573**	**271,567**	**44,245**	**33,764**	**130**	**118**

S Data do not meet publication standards due to high sampling variability or other reasons. [1] Truck as a single mode includes shipments that went by private truck only, for-hire truck only, or a combination of private truck and for-hire truck. [2] Commodity Flow Survey data exclude shipments of crude oil.

Source: U.S. Department of Transportation, Research and Innovative Technology Administration, Bureau of Transportation Statistics, and U.S. Department of Commerce, U.S. Census Bureau, 2007 Commodity Flow Survey, <http://factfinder.census.gov/>, accessed April 2011.

Table 1071. Hazardous Shipments—Value, Tons, and Ton-Miles: 2002 and 2007

[660,181 represents $660,181,000,000. For business establishments in mining, manufacturing, wholesale trade, and selected retail industries. 2007 industries classified by the 2002 North American Industry Classification (NAICS). 2002 industries classified by the 1997 North American Industry Classification. Selected auxiliary establishments are also included. Based on the 2007 Economic Census; see Appendix III]

Figure 7.1 Shipment characteristics by mode of transportation

The following subsections cover each mode of transport in more detail.

Rail

Rail is the slowest, least flexible, yet lowest-cost mode of transportation. So, it is typically used to transport bulky commodities over long distances. Because rail carriers must provide own ways, they are actually *natural monopolies*, but still must provide their own terminals and vehicles, resulting in a large capital investment and high volumes required to operate a railroad. As a result, they tend to have high fixed costs and low variable costs.

Railroads come in three general types:

- **Class I:** At least 350 miles in track and/or revenue at least $272 million (in 2002 dollars). Class I carriers comprise only 1% of the number of U.S. freight railroads, but they account for 70% of the industry's mileage operated, 89% of its employees, and

92% of its freight revenue. Class I carriers typically operate in many different states and concentrate largely (though not exclusively) on long-haul, high-density intercity traffic lanes. There are seven Class I railroads: BNS, Canadian National, Canadian Pacific, CSX Transportation, Kansas City Southern, Norfolk Southern, and Union Pacific.

- **Regional and local line haul:** Regional railroads are line-haul railroads with at least 350 route miles and/or revenue of between $40 million and the Class I threshold. There were 31 regional railroads in 2002. Regional railroads typically operate 400 to 650 miles of road serving a region located in two to four states.

  Local line haul carriers operate less than 350 miles and earn less than $40 million per year. In 2002, there were 309 local line haul carriers. They generally perform point-to-point service over short distances. Most operate less than 50 miles of road (more than 20% operate 15 or fewer miles) and serve a single state.

- **Switching and terminal (S&T) carriers:** Railroads that primarily provide switching/terminal services, regardless of revenue. They perform pickup and delivery services within a certain area. In 2002, there were 205 S&T carriers. The largest S&T carriers handle hundreds of thousands of carloads per year and earn tens of millions of dollars in revenue (Association of American Railroads, 2004).

Motor Carriers

Motor carrier is the most widely used mode of transportation in the domestic supply chain; most consumer products are shipped via this method from manufacturers, wholesalers, and distributors to retailers. In fact, there are more than half a million private, for hire, and other U.S. interstate motor carriers.

The economic structure of the motor carrier industry contributes to the vast number of carriers in the industry because it has low fixed and high variable costs.

Motor carriers pay for highway, tunnel, and bridge access through taxes or tolls and provide their own terminals and are fairly fast and flexible because they can offer door-to-door service and are used for small-volume goods to many delivery locations.

Within this mode, there are full-truckload, less-than-truckload (LTL; national and regional), and small-package carriers. Examples of for-hire carriers include Schneider and Werner (TL), Con-way Freight and Old Dominion Freight Line (LTL), and UPS (small package).

Full-truckload carriers have the lowest overhead because they gain economies by filling out one load or trailer with one customer's freight and go *point to point*. In contrast, LTL and small-package carriers must have a network of terminals called *break bulks* for sorting and mixing as each vehicle may have dozens of customer's small shipments going to a variety of places. This infrastructure is reflected in their rates, as discussed later in the chapter.

Air Carriers

Air cargo carriers are the fastest, most expensive mode of transportation and are an especially important part of many international logistics networks. They use government-provided terminals and air traffic control systems, so they have relatively low fixed costs but operate with high variable costs for fuel and operating costs and tend to haul high-value, lower-volume, and time-sensitive cargo at premium rates.

Some cargo airlines are divisions or subsidiaries of larger passenger airlines; others, such as UPS, FedEx, and DHL, operate for cargo only.

Water Carriers

As noted earlier, water transport is one of the oldest forms of transport. It is divided between domestic and (deepwater) international transport.

Nature provides ways in most cases; however, canals and ports are government controlled. The carrier pays for use of terminals and owns the ships, so there are moderate fixed costs but low operating costs.

This mode is fairly slow and not very flexible and is used to haul low-value bulk cargo over long distances. However, with the advent of containerization in the 1970s, it has become a major facilitator of international trade, carrying 81% international freight movement.

Intermodal Carriers

Intermodal refers to freight being transported in an intermodal container or vehicle. The most widely used intermodal systems are the trailer on a flatcar (TOFC) and container on a rail flatcar (COFC).

This takes advantage of the economies of each mode of transportation (rail, ship, and truck), with no handling of the freight itself when changing modes. As a result, this improves accessibility, reduces cargo handling, and improves security. It also reduces damage and loss and increases the speed with which freight is transported.

It has also helped to facilitate the growth of global trade when used in conjunction with water transport on container ships with standardized containers that are compatible with multiple modes of transport.

Pipeline

Pipelines are a unique mode of transportation used for high-volume gases or liquids moving from point to point. The equipment is fixed in place, and the product moves through it in high volume. There are 174 operators of hazardous liquid pipelines that primarily carry crude oil and petroleum products, the most well known of which is the Trans-Alaska Pipeline System (TAPS).

Crude oil pipelines are the basis for our liquid energy supply. The crude oil is collected by pipelines from inland production areas like Texas, Louisiana, Alaska, and western Canada. Pipelines also move crude oil produced far offshore in coastal waters as well as from Mexico, Africa and the Middle East, and South America delivered by marine tankers, often moving for the final leg of that trip from a U.S. port to a refinery by pipeline.

In addition, two-thirds of the lower 48 states are almost totally dependent on the interstate pipeline system for their supplies of natural gas.

Global Intermediaries

There also exists as many *global intermediaries* as there are a variety of services required for international trade. Some of them are as follows:

- **Freight brokers:** Similar to any other type of broker, the main function is to bring together a buyer and a seller. The buyer in this case is the shipper of the goods, and the seller is the trucking company. The broker negotiates the terms of the deal and handles much of the paperwork.

- **Freight forwarders:** Heavily utilized in global trade, for both surface and air, to comply with export documentation and shipping requirements, many exporters utilize a freight forwarder to act as their shipping agent. The forwarder advises and assists clients on how to move goods most efficiently from one destination to another. A forwarder has extensive knowledge of documentation requirements, regulations, transportation costs, and banking practices, thus assisting in the exporting process for many companies.

  They may also provide essential freight services such as assembling and consolidation of smaller shipments plus taking larger bulk shipments and breaking them into smaller shipments.

- **Customs brokers:** Perform transactions at ports on for other parties. Typically, an importer hires a customs broker to guide their goods into a country. Similar to the forwarder, the broker will recommend efficient means for clearing goods through customs entry and can also estimate the landed costs for shipments entering the country. U.S. exporters typically do not book shipments directly with a foreign customs broker, because freight forwarders often partner with customs brokers overseas who will clear goods that the forwarder ships to the overseas port. However, foreign customs brokers contract the services of the domestic freight forwarder when the goods are headed in the opposite direction.

  The types of transactions negotiated for an importer may include the entry of goods into a customs territory, payment of taxes and duties, and duty drawback or refunds of any kind.

- **Non-vessel-operating common carriers (NVOCCs):** NVOCCs are also freight forwarders, except that they 1) may own and operate and sometimes lease the containers they ship, 2) be required to publish a public tariff (that is, rates), 3) may have to take on the status of a *virtual carrier* and take on liabilities of a carrier, and 4) whereas freight forwarders can be agents for an NVOCC, the reverse is not true, giving NVOCCs more flexibility.

Legal Types of Carriage

There are two legal types of carriers, for hire and private, as described here.

For Hire

For-hire carriers offer service to the general public and are subject to government regulations in regard to rates, routes, and markets served.

They come in two major forms:

- **Common carriers:** Licensed to carry only certain goods available to public to designated points or areas served and offer scheduled service. Common carriers must file both liability insurance and cargo insurance. Public airlines, railroads, bus lines, taxicab companies, cruise ships, motor carriers, and other freight companies generally operate as common carriers (as do communications service providers and public utilities).
- **Contract carriers:** For-hire interstate operators that offer transportation services to certain shippers under contracts. Contract carriers must file only liability insurance.

There are also independent carriers, referring to an *individual owner-operator* or *trucker* who can make agreements with private carriers, common carriers, contract carriers, or others as they want.

Private

Carriers are considered private when a company transports only their own goods. Their primary business is not transportation, and the vehicles are not for hire. Private carriage usually refers to trucking, but is also found in rail and water transportation.

Very high volume or specific needs are needed to justify the expense. Many retail organizations, as well as some manufacturers, distributors, and wholesalers, operate their own fleets. We've all seen Walmart and Toys R Us trucks (with Jeffrey Giraffe) printed on the side of the trailer on a highway at one time or another. These are examples of private carriage.

In many cases, private vehicles such as those mentioned here can also be used to *backhaul* freight from suppliers after delivering product from distribution facilities to retail locations.

This can avoid the need for the vehicle to make the return trip empty and to reduce their fleets' overall operating costs.

Transportation Economics

In this section, we will deal with the application of demand and cost principles to transportation.

Transportation Cost Factors and Elements

There are a variety of factors that impact transportation costs, which we will now cover.

Cost Factors

The primary factors influencing transportation costs pricing are distance, weight, and density (see Figure 7.2a, b, c).

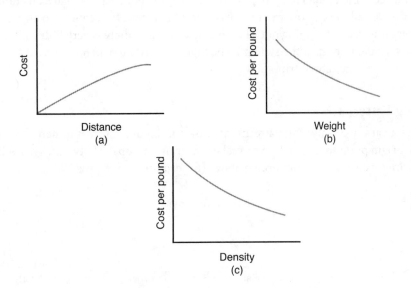

Figure 7.2 Transportation economics

- **Distance:** As distance traveled increases, variable expenses such as labor, fuel, and maintenance increase.
- **Weight:** The transport cost per unit decreases as the load weight increases. This is as a result of *economies of scale* as a result of spreading the fixed costs over more weight. That is why it is always best to combine small loads into larger ones where possible, up to the capacity of the vehicle carrying the load.

- **Density:** This is the combination of weight and cubic volume. Vehicles have both weight and cubic capacity and tend to *cube out* before *weighting out*. So, shipping higher-density products enables the cost to be spread over more weight versus a lighter load such as paper cups, where you are shipping a lot of air for your money. As a result, higher-density products are charged less per *hundredweight* or *CWT* (that is, per hundred pounds; one of the common forms of transportation pricing, which is discussed shortly).

- **Stowability:** Similar to density and is also a factor. It refers more to the ease of storage, such as when shipping items are easily stacked or nested.

- **Other factors:** Can includes factors such as the amount of handling necessary (smaller volumes typically require more handling), type of handling (full pallets can be handled with more automated equipment versus individual cases, which are handled manually in most cases), liability (can be a function of the value or nature of the product), perishability, and market factors such as market (origin and destination) volume and balance, which refers to amount of freight flowing both ways. Finally, an empty return or backhaul can be costly to the carrier and thus may affect rates into an area that has very little freight coming out. (For example, because there is very little manufacturing in Florida, rates into the state may be high because carriers have a hard time finding freight for the return trip.)

Shipping Patterns

There are a variety of shipping patterns, as shown in Figure 7.3. A shipment may go direct from the origin point to destination or make one or more stops in between. This will depend on the primary cost factors mentioned above (for example, TL versus LTL).

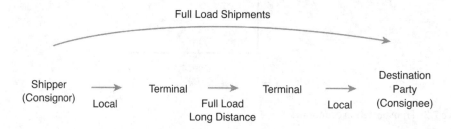

Figure 7.3 Common shipping patterns

Cost Elements

Four major elements, the principles, are the same for all modes:

- **Line haul:** Carriers have basic costs to move product from point to point that include fuel, labor, and depreciation. The costs are pretty much the same per mile whether the container is full or empty, so the line haul cost is total of these costs divided by the distance traveled.

  So, for example, if the line haul cost to transport material from point A to point B is $5 per mile and the route is 500 miles, the total line haul cost is $2,500. The actual cost (in $/CWT) for a shipment weighing 12,000 pounds is $20.83/CWT (that is, $2,500 / 120), and the cost for a shipment weighing 45,000 pounds is $5.56/CWT.

  As a result, the total line haul cost will vary with the cost *per mile* to operate the vehicle and the distance the material is moved. The line haul cost *per hundredweight* varies based on the cost per mile, distance, and weight, as you learned in the earlier example.

Other line haul services may be required, as well, including the following:

- **Reconsignment:** This involves changing the consignee while the shipment is in transit and is used commonly in industries where goods are shipped before they are sold.

- **Diversion:** The changing of the destination of a shipment while in transit, which is often used in conjunction with reconsignment.

- **Pooling:** This allows a shipper to use a less-costly container or truckload rate by consolidating smaller shipments going to one destination and one consignee into one pool car or truck.

- **Stopping in transit:** This allows the shipper to use a full container or truckload rate and drop off portions of the load at various intermediate destinations. The shipper is invoiced for a stop-off charge for each stop, which is usually a lot less than shipping the load at less than car or truck load rates.

- **Transit privilege:** This allows for the shipper to unload a car or trailer, process the shipment, and then reload and ship the processed product to its final destination using a through rate (that is, a single transportation rate on an interline haul made up of two or more separately established rates).

- **Pickup and delivery:** These costs depend on the time spent picking up and dropping off cargo and not distance. There is a charge for each pickup, so it is useful to consolidate multiple shipments to avoid multiple separate trips.

  The loading and unloading of freight at pickup and delivery is generally the responsibility of the carrier in the case of LTL or LCL (and small-package) shipments, whereas the shipper is usually responsible for TL and CL loading and unloading.

The carrier will specify the amount of time the shipper and receiver have for loading and unloading. In the case of rail *free time*, this is 24 to 48 hours; after free time, rail carriers charge what is called a *demurrage fee*; motor charge what is called a *detention fee*. Motor carrier loading and unloading times vary, but can be as little as a half hour.

- **Terminal handling:** These costs depend on how many times the shipment must be handled. In the case of full truckloads (TL), there is no terminal handling because they go directly to the customer. However, LTL shipments must be sent to a terminal, sorted, and consolidated, so charges are incurred. As a result, it is wise to consolidate shipments into fewer parcels where possible.

There are other terminal services besides handling, including the following:

- **Consolidation:** Many small shipments are consolidated into a one larger shipment going to a customer, qualifying the shipper for a lower rate.
- **Dispersion:** This is the opposite of consolidation, with one large shipment being distributed to multiple customers at the destination terminal.
- **Shipment services:** The carrier provides freight handling for consolidation or dispersion.
- **Vehicle service:** Carriers need to maintain a diverse and adequate fleet of transit vehicles for shipper's use.
- **Interchange:** Carriers must provide the ability to interconnect with other carriers of the same or different modes so that through rates may be used by the shipper.
- **Weighing:** The carrier (or shipper) provides the weight of shipment.
- **Tracing:** Carriers can communicate to shipper where the shipment is and when it might be delivered. In most cases, this information can be supplied via the Internet.
- **Expediting:** In some cases, it is necessary to move a shipment faster than normal, and as a result, this may require a premium over-regular handling.
- **Billing and collecting:** This includes the costs of paperwork and invoicing the shipper. Carriers also provide clerical services for bills of lading (documents issued by a carrier for a shipment of merchandise giving title of that shipment to a specified party), freight bills, and routing of the shipment.

Rates Charged

Now that we you understand the general economics of transportation cost and it its major elements, let's turn our focus to pricing.

Effects of Deregulation on Pricing

Since deregulation, transportation rates or prices are negotiated like other commodities. Some changes as a result of deregulation by mode are as follows:

- **Motor and water carriers:** Rate and tariff-filing regulations were eliminated except for household and noncontiguous trade for domestic water transportation (that is, shipments that originate or terminate in Alaska, Hawaii, or a U.S. territory or possession). The common carriage concept was for the most part eliminated, as all carriers may contract with shippers. Antitrust immunity for collective ratemaking was eliminated.

- **Air carriers:** As mentioned earlier, economic regulation of air carriers was eliminated in 1977. However, safety regulation remains in force.

- **Rail carriers:** This is still the most regulated of the transportation modes. There has been, however, complete deregulation over certain types of traffic: piggyback and fresh fruits.

- **Freight forwarders and brokers:** Both are required to register with the Surface Transportation Board (STB). Brokers must also post a $75,000 bond to ensure payment to the carriers. No economic rate or service controls these entities. A freight forwarder is considered a carrier and is liable for freight damages.

Pricing Specifics

Full container or truck load rates may be expressed in a flat dollar or mileage rate, and less than containers or truckloads may be a discount off of the class rate from the tariff.

In general, prices (known as *rates*) are expressed in either dollars (whole or per mile) or cents per hundredweight (CWT) and are contained in tariffs, which can be in hard copy or electronic form.

Freight Classifications

The classification of an item must first be determined. The classification is based on the cost elements of an item mentioned earlier: density, stowability, ease of handling, and liability. The class given to an item is known as its *rating*.

Truck and rail each have their own set of classifications. For motor carriers, it is the National Motor Freight Classification (NMFC), and for rail, it is the Uniform Freight Classification (UFC).

A class of 100 is considered average. A class can range from 35 to 500. (In general, the higher the rating, the higher the transportation cost.)

Rate Determination

After the class is identified, the rate must be determined and is based on the origin and destination. There is usually a minimum charge and various rates at *weight breaks* as the shipments increase in size. There may also be rate surcharges or accessorial charges for extra services provided by the carrier.

There are also other types of rates, such as the following:

- **Cube or density rates:** Freight rate computed on the basis of a cargo's volume, instead of its weight.
- **Exception rates:** A deviation from the class rate; changes (exceptions) made to the classification.
- **Commodity rates:** The carrier will offer an all-commodity rate for this specific route despite the class of the commodity carried. The class of the commodity does not matter to the carrier.
- **Freight-all-kinds (FAK) rates:** These are rates for a carrier's tariff classification for various kinds of goods that are pooled and shipped together at one freight rate. Consolidated shipments are generally classified as FAK.

Documents

A variety of documents are used in transportation both domestically and internationally. This section covers the main ones in this section.

Domestic Transportation Documents

Before discussing the major domestic documents, it is first useful to understand *terms of sale*.

Terms of Sale

Transportation costs are the second- or third-highest expense that a manufacturing company has beyond the cost of labor and raw materials, so it makes sense to know how they are allocated. Even if your vendor pays the freight charges, you need to know the amount they paid, because at some point when you have enough volume, you may want to take control of your inbound freight and negotiate rates with your own carriers to less than you are paying now. You should always identify freight costs separate from cost of goods.

Negotiating the most appropriate terms of sale will allow you to add value to your purchase.

The terms determine which party is to pay the freight bill, which party has title to the goods, and which party controls the movement of the goods.

The two major terms are as follows:

- **F.O.B. origin:** The buyer pays for the freight, takes title to the goods once loaded, and controls movement of the goods.
- **F.O.B. destination:** The seller pays for the freight, has title to the goods until they are delivered, and controls movement of the goods.

There are variations to these terms, as follows:

- **F.O.B. origin, freight collect:** The buyer pays freight charges, owns goods in transit, and files claims, if any.
- **F.O.B. origin, freight prepaid:** The seller pays freight charges, and the buyer owns goods in transit and files claims, if any.
- **F.O.B. origin, freight prepaid and charged back:** The seller pays freight charges, owns goods in transit, and the buyer files claims, if any.
- **F.O.B. destination, freight collect:** The buyer pays freight charges, and the seller owns goods in transit and files claims, if any.
- **F.O.B. destination, freight prepaid:** The seller pays freight charges, owns goods in transit, and files claims, if any.
- **F.O.B. destination, freight collect and allowed:** The buyer pays freight charges, and the seller owns goods in transit and files claims, if any.

Bill of Lading

A bill of lading (B/L) is a contract between the carrier and the shipper issued by a carrier, which details a shipment of merchandise and gives title of that shipment to a specified party (that is, a receipt) with specified timing.

A B/L includes title to the goods and name and address of the consignor and consignee and summarizes the goods in transit and their class rates. Electronic bills are now used often where the carrier and shipper have an established strategic partnership.

There are two main types of B/Ls:

- **(Uniform) straight bill of lading:** These are nonnegotiable and contain terms of the sale, including the time and place of title transfer.
- **Order (notified) bill of lading:** These are negotiable, and the consignor retains the original until the bill is paid. They can be used as a credit instrument because there is no delivery unless the original bill of lading is surrendered to the carrier.

There are also export bills of lading (covered in the section "International Transportation Documents") and government bills of lading. Government B/Ls are used when the product is owned by the U.S. government.

In cases where there are individual stops or consignees when multiple shipments are placed on a single vehicle, what is known as a *shipment manifest* is used. Each shipment still requires a B/L, and the manifest lists the stop, B/L, weight, and case count for each shipment. The goal of a manifest is to provide one document that describes the complete contents of the load.

The B/L also documents responsibilities for all possible causes of loss or damage and includes terms such as the following:

- Common carrier liable for all losses, damage, or delays in shipment.
- Exceptions include acts of God, public enemy, shipper, public authority, and inherent nature of the goods.
- Reasonable dispatch.
- Shipper liable for mending, cooperage, bailing, or reconditioning of goods or packages and gathering of loose contents for packages.
- Freight not accepted stored at owner's cost.

Freight Bills

Freight bills are the carrier's invoice for charges for a given shipment. The credit terms are specified by the carrier and can vary extensively. In some cases, credit may be denied if the charges are worth more than the freight.

Bills may also be either prepaid or collect per the previous discussion on freight terms.

Because there tends to be large changes to fuel costs, low visibility of the future freight costs, and a relatively high complexity of freight quotes, freight invoices are susceptible to human and process errors and require auditing to ensure that the organization does not overpay for services it did not incur.

These audits can be performed internally or externally, both prepayment and, in some cases, postpayment.

When I worked in General Electric Corporate Sourcing, I helped to establish the GE Corporate Freight Payment Center in Fort Myers, Florida. The two major goals of the service was to both consolidate information for their 100 or so business units to leverage the over $1 billion spent company-wide on transportation services and to perform a pre-audit on freight bills in a more standardized, automated fashion (because freight bill errors, including overcharges and duplicate bills and payments, can range as high as 5% domestically and even as high as 10% internationally).

Freight Claims

A freight claim is a document filed with the carrier to recover monetary losses due to losses, damage, delay, or overcharges by the carrier. In most cases, claims are filed within 9 months, the claimant is notified by receipt within 30 days, and settlement or refusal usually occurs within 120 days.

The claimants are expected to take some reasonable measures to minimize the loss, such as requiring the carrier to pay for the difference between the original value and the damaged or salvage value.

International Transportation Documents

By its very nature, documentation for international transportation is much more complex than required for domestic transportation. The types of documents vary widely by country and fall into two major categories of sales and transportation.

Sales Documents

A *sales contract* is usually the initial document used international trade. A *letter of credit*, a document issued by a financial institution ensuring payment to a seller of goods or services, may also accompany shipment.

For export, one may need an *export license*, which is the express authorization by a country's government to export a specific product before it is shipped. Governments may require an export license to exert some control over foreign trade for political or military reasons, control the export of natural resources, or control the export of national treasures or antiques.

Also for export, a shipper's *export declaration* is required by U.S. Customs, which is designed to keep track of the type of goods exported from the United States, as well as their destination and their value.

There may also be *export taxes* and *quotas* in effect.

For import, countries require certain documents to ensure that no shoddy quality goods are imported; and to help determine the appropriate tariff classification, the correct value of imported goods, the correct country of origin for tariff purposes; or to protect importers from fraudulent exporters or limit (or eliminate) imports of products that the government finds inappropriate for whatever reason.

Import documents include the following:

- **Certificate of origin:** A document provided by the exporter's chamber of commerce that attests that the goods originated from the country in which the exporter is located. It is used by the importing country to determine tariff of goods.

- **Certificate of manufacture:** A document provided by the exporter's chamber of commerce that attests that the goods were manufactured in the country in which the exporter is located.

- **Certificate of inspection:** A document provided by an independent inspection company that attests that the goods conform to the description contained in the invoice provided by the exporter and that the value of the goods is reflected accurately on the invoice. It is always obtained by the exporter in the exporting country, before the international voyage takes place, and the certificate of inspection is the result of a pre-shipment inspection (PSI).

- **Certificate of free sale:** This shows that the goods sold by the exporter can legally be sold in the country of export; this certificate is designed to prevent the export of products that would be considered defective in the country of export.

- **Import license:** A document issued by the importing country that is designed to prevent import of nonessential or overly luxurious products in developing countries short of foreign currency supply.

- **Certificate of insurance:** Some international terms of sale, or *Incoterms* (International Commerce Terms), require that the exporter provide insurance, and a certificate of insurance offers this proof of coverage.

- **Carnet:** International customs documents that simplify customs procedures for the temporary importation of various types of goods. They ease the temporary importation of commercial samples, professional equipment, and goods for exhibitions and fairs by avoiding extensive customs procedures and eliminating payment of duties and value-added taxes (minimum 20% in Europe, 27% in China); they replace the purchase of temporary import bonds.

Terms of Sale

International Commercial Terms, also known as *Incoterms*, are a set of rules that define the responsibilities of sellers and buyers for the delivery of goods under sales contracts for domestic and international trade. They are published by the International Chamber of Commerce (ICC) and are widely used in international commercial transactions. They provide a common set of rules to apportion transportation costs and clarify responsibilities of sellers and buyers for the delivery of goods under sales contracts. The goal is to simplify the drafting of contracts and help avoid misunderstandings by clearly describing the obligations of buyers and sellers.

The terms may include export packing costs, inland transportation, export clearance, vehicle loading, transportation costs, insurance, and duties.

The two main categories of Incoterms® 2010 are now organized by modes of transport:

- Group 1: Incoterms* that apply to any mode of transport.
  - EXW Ex Works
  - FCA Free Carrier
  - CPT Carriage Paid To
  - CIP Carriage and Insurance Paid To
  - DAT Delivered at Terminal
  - DAP Delivered at Place
  - DDP Delivered Duty Paid
- Group 2: Incoterms* that apply to sea and inland waterway transport only.
  - FAS Free Alongside Ship
  - FOB Free on Board
  - CFR Cost and Freight
  - CIF Cost, Insurance, and Freight

International Transportation Documents

Transport documents are a crucial part of international trade transactions. The documents are issued by the shipping line, airline, international trucking company, railroad, freight forwarder, or logistics companies.

To the shipping company and freight forwarder, transport documents provide an accounting record of the transaction, instructions on where and how to ship the goods, and a statement giving instructions for handling the shipment.

International Bill of Lading

B/Ls in international trade help guarantee that exporters receive payment and that importers receive merchandise. The export B/L can govern foreign,, intercountry, and domestic movements of the goods.

An international B/L can have a number of additional attributes, such as on-board, received-for-shipment, clean, and foul. An on-board B/L denotes that merchandise has been physically loaded onto a shipping vessel, such as a freighter or cargo plane. A received-for-shipment B/L denotes that merchandise has been received, but is not guaranteed to have already been loaded onto a shipping vessel. Such bills can be converted upon being loaded.

Some of the specific types of B/Ls related to international transportation include the following. (Straight and intermodal for domestic have already been discussed.)

- **Ocean bill of lading:** Sets terms and lists origin and destination ports, quantities and weight, rates, and special handling needs for the ocean movement. The ocean carrier is held liable for losses due to negligence only, with other losses being the responsibility of the shipper. A commercial invoice determines the value of the products in the case of losses due to negligence.

- **Air waybill:** A B/L used in the transportation of goods by air, domestically or internationally.

B/Ls can also be for receipt or title to goods in the form of a

- **Negotiable or to order B/L:** A negotiable B/L allows the owner of the goods to sell them while they are in international transit. The transfer of ownership to the new owner is done with the B/L, because it is a certificate of title to the goods. Only ocean B/Ls can be negotiable.

- **Clean B/L:** This type of B/L is used as a receipt for goods and is issued by carrier when goods arrive in port; damages and other exceptions should be noted. (A *foul* B/L denotes that merchandise has incurred damage prior to being received by the shipping carrier. Letters of credit usually will not allow for foul B/Ls.)

Key Metrics

In addition to budgeting transportation costs by mode and by lane, a variety of performance measurements are used in transportation for current performance versus historical results, internal goals, and carrier commitments. The main categories of key metrics are service quality and efficiency and may include on-time delivery, loss and damage rate, billing accuracy, equipment condition, and customer service.

Technology

Transportation management systems (TMSs) have been around for a long while. Historically, they have been an *add-on* to an existing enterprise resource planning (ERP) or legacy (that is, homegrown) order-processing or warehouse management system (WMS).

Like most software today, they can be installed as resident software or web based and accessed on demand.

A TMS offers benefits to an organization such as automated auditing and billing, optimized operations, and improved visibility.

They typically include functionality to plan, schedule, and control an organization's transportation system, with functionality for the following:

- **Planning and decision making:** Helps to define the most efficient transport schemes according to parameters such as the following: transportation cost, lead time, stops, and so on. Also includes inbound and outbound transportation mode and transportation provider selection and vehicle load and route optimization.

- **Transportation execution:** Allows for the execution of a transportation plan such as carrier rate acceptance, carrier dispatching, electronic data interchange (EDI), and so on.

- **Transport follow-up:** Tracking of physical or administrative transportation operations such as traceability of transport event by receipt, custom clearance, invoicing and booking documents, and sending of transport alerts (delay, accident, and so on).

- **Measurement:** Cost control and key performance indicator (KPI) reporting as it relates to transportation.

Ultimately, a supply chain system is made up of connecting links and nodes, where the transportation system provides the *links*, and the facilities provide the *nodes*. Therefore, the next logical topic to cover is warehouse management and operations.

8

Warehouse Management and Operations

A warehouse is typically viewed as a place to store inventory. However, in many supply chain systems, the role of the warehouse is more properly viewed as a switching facility versus just a storage facility because it provides time and place utility for raw materials, industrial goods, and finished products, allowing firms to use customer service as a value-adding competitive tool.

Therefore, warehouses are one of the more complex areas of the supply chain because they are the point of intersection for all the members of the supply chain.

Even the name is a bit of a misnomer. The term *warehouse* implies that something is stored or *warehoused* for a fairly long time. That might be true in many cases, but in today's faster-paced world, material tends to fly though warehouses rather than stay put long (that is, if you want to remain competitive and stay in business for the long run).

More often than not, on the outbound end (typically finished goods or components), the facilities are referred to as *distribution centers* (DCs), whereas *warehouse* is more often used to describe a facility where raw materials and parts are stored.

Still others such as e-commerce businesses or e-tailers refer to their facilities as *warehouse and fulfillment centers* or variations on that, because order fulfillment is the complete process from point-of-sales inquiry to delivery of a product to the customer. Sometimes order fulfillment is used to simply describe the act of distribution or the shipping function as per the above, but in the broader sense it refers to the way firms respond to customer orders and to the process they take to move products from those orders to the customer.

For our purposes, we define the facilities as follows:

- **General warehouse:** This type of facility is primarily for the storage and protection of goods, with the need to minimize handling and movement.
- **Distribution warehouse (or distribution center):** In these facilities, goods are received in large volumes. The goods are then sorted, stored, and then consolidated into customer orders for fulfillment. This type of facility is more concerned with throughput.

For simplicity, we will just refer to them as warehouses or DCs from now on in this chapter.

Like transportation, warehousing offers a variety of professional career opportunities at the corporate, operations, and sales areas. In corporate, there are positions that oversee both private and public warehouses from both a strategic and operational perspective.

In the warehouses themselves, positions range from supervisors, analysts, and managers to general managers of facilities. In addition, in public warehousing, there are a variety of sales positions. There is also an organization known as WERC (Warehouse Educational and Research Council; www.werc.org) for professional networking, education, and career searches.

Brief History of Warehousing in America

The original warehouses in the United States were sheds at docks. Once the railroad system was established in the mid-1800s, warehouses were established for consolidation and distribution to support it.

As the transportation system further developed, warehousing shifted more to manufacturers, wholesalers, and retailers.

After WWII, as the retail industry grew rapidly, it became necessary for retailers to establish DCs to provide a better assortment of products to consumers as well as to gain transportation and purchasing economies.

Today, warehouses and DCs may be highly automated as inventory moves quickly through them to support JIT (just-in-time) production processes and the consumers' ever-increasing demand for a vast assortment of products in stores and fast delivery to their homes.

#60148575 ©industrieblick-Fotolia.com

Economic Needs for Warehousing

We need warehousing for a variety of reasons, such as the following:

- **Seasonal production:** There is a need for proper storage or warehousing for products such as agricultural commodities, from where they can be supplied as and when required.

- **Seasonal demand:** Some goods are demanded seasonally like snow blowers in the winter or lawnmowers in the summer. The production of these goods may take place throughout the year to meet the peak seasonal demand. As a result, there is a need to store these goods in a warehouse to make them available at the time of need.

- **Production economies of scale:** Manufacturers typically produce in lots or batches to meet existing as well as future demand of the products to enjoy the benefits of the *economies of scale* of mass production. These large quantities of finished products, produced on a large scale, need to be stored properly until purchased by the customer.

- **Quick supply:** Goods of all types are produced at specific sites but consumed throughout the country and perhaps the world. Because the manufacturing sites may be far from the ultimate consumer, they may be supplied via a large, distant distribution network. Therefore, it is essential to stock these goods near the place of consumption so that without any delay, these goods are made available to the consumers at the time of their need.

- **Continuous production:** The continuous production of goods in factories requires an adequate supply of raw materials. As a result, there is a need to keep sufficient quantities of raw material in the warehouse to ensure continuous production.

- **Price stabilization:** To keep the price of the goods in the market stable, there is a need to keep sufficient stock in warehouses because scarcity in supply of goods may increase their price in the market. Conversely, excess production and supply may also lead to a fall in prices of the product. So, by keeping the supply of goods in balance, warehousing can contribute to price stabilization.

Types of Warehouses

We will now look at various types of warehouses that exist today from a number of views. One view is by customer classification, another by role in the supply chain, and yet another is by ownership type.

Warehouses by Customer Classification

When thinking of warehouses by customer classification, we come up with the following types:

- **Factory warehouse:** This type of facility connects production with wholesalers. It typically supplies a small number of large orders daily with advance information about order makeup.

- **Retail distribution warehouse:** These serve a number of internal retail units. They have advance information about order detail and perform carton and item picking from a forward area. They generate more orders per shift than DCs that perform consolidation.

- **Catalog retailer/e-tailer:** This is a fulfillment warehouse filling orders from catalog sales a large number of small (frequently single-line) orders. There is item and, sometimes, carton picking daily with advance composition of orders usually unknown (only statistical information available).

- **Support of manufacturing operations:** This is usually a relatively small warehouse or stock room that provides raw material and/or work in process to manufacturing operations. There are many small orders driven by production schedules and, in some cases (parts, hardware, and so on), there is only statistical information available about order composition. There are usually stringent, short time requirements.

Warehouses by Role in the Supply Chain

Since World War II, as the consumer market grew, the function of warehousing has evolved from pure storage to more speed, cost reduction, flexibility, and efficiency. This has led to warehouses that are more geared to this, and they include the following:

- **Distribution centers (DCs):** This type of facility is stocked with products (usually finished goods) to be redistributed to retailers, wholesalers, or directly to consumers. DCs are usually thought of as being demand driven, and are often referred to as *retail distribution centers*, per above, when they primarily distribute goods to retail stores, *order-fulfillment* center commonly when they distribute goods directly to consumers, and *cross-dock* facilities (see later in this list) when they store little or no product but distribute goods to other destinations.

- **Consolidation:** These facilities receive materials from a number of sources and combine them into exact quantities for a specific destination. A local or central DC may play this role.

- **Break-bulk:** This is when a warehouse receives a single large shipment and arranges for delivery to multiple destinations. Break-bulks are common in the less-than-truck-load (LTL) trucking industry.

- **Cross-docking:** There is also a type of warehouse (or section of a warehouse) used heavily in retail distribution to deliver inventory to retail locations known as *cross-dock*. This entails the practice of unloading materials from an incoming semi-trailer truck or railroad car and loading these materials directly into outbound trucks, trailers, or rail cars, with little or no storage in between. It is common for inventory to flow through this type of facility within 24 to 48 hours. Cross-docking has become an integral part of the supply chain strategy of retailers such as Walmart and Toys R Us to turn inventory rapidly in the DCs and to resupply stores rapidly.

- **Reverse logistics:** Whereas many warehouses have a small section of their facility to handle returns, some are entirely dedicated to what is known as *reverse logistics* for items that are going from the end user back to the distributor or manufacturer. Specifically, this is the process of moving goods from their final destination for the purpose of capturing value, or proper disposal, and may also include remanufacturing and refurbishing activities.

Warehouses by Ownership Type

There are a number of general categories of warehouses, as follows:

- **Public warehouse:** The public warehouse is essentially space that can be leased to solve short- or long-term distribution needs. Public warehouse fees can be a combination of storage fees and inbound and outbound transaction fees. A public warehouse can charge per pallet or charge for each square foot that is used by a company. Retailers that operate their own private warehouses may occasionally seek additional storage space if their facilities have reached capacity or if they are making a special, large purchase of products. For example, retailers may order extra merchandise to prepare for in-store sales or order a large volume of a product that is offered at a low promotional price by a supplier. Their use varies by industry, but it is common in the household product, personal-care, and grocery industries to use shared public warehouse facilities. Some of these public warehouses also provide value-added services such as light assembly and kitting and may operate their own private fleet of trucks to facilitate local delivery.

  Fees for public warehousing are both time based (that is, storage charges) and transaction based (that is, handling charges; in/out handling fees, documentation, special services, and so on).

- **Private warehouse:** This type of warehouse is owned and operated by channel suppliers and resellers and used in their own distribution activity. For instance, a major retail chain may have several regional warehouses supplying their stores, or a wholesaler will operate a warehouse at which it receives and distributes products. Private warehouses require a capital investment to be made by their owner, but these warehouses can be extremely cost-effective in the long run.

- **Contract warehouse:** As the name implies, a contract warehouse handles shipping, receiving, and storage of products on a contract basis. Contract warehouses generally require a client to commit to a specific period of time (generally in years) for the services. Contracts may require clients to purchase or pay for storage and material-handling equipment. Fees for contract warehouses may be transaction and storage based, fixed, cost plus, or any combination.

- **Bonded warehouse:** A bonded warehouse is licensed to accept imported goods for storage before payment of customs duties. By storing his goods in a bonded warehouse, the importer gains some control without paying the duty. The goods in bonded warehouses are under the supervision of customs officers, and permission is necessary before the owner can access them.

- **Government warehouse:** These warehouses are owned, managed, and controlled by central or state governments, public corporations, or local authorities. Both government and private enterprises may use these warehouses to store their goods.

  In many cases, this type of warehouse is mainly located at the important sea ports, and in most cases is owned by the dock authorities. The general public can also use this group of warehouse on payment of fixed charges. If a customer cannot pay the rent within the specified time or period, the authority can recover the rent by disposing of the goods.

- **Co-operative warehouse:** These warehouses are owned, managed, and controlled by co-operative societies. They provide warehousing facilities at the most economic rates to the members of their society.

Warehouse Features

All the previously described types of warehouses, depending on the industry they serve, may have features that define them, such as the following:

- **Automated warehouses:** With advances in computer and robotics technology, many warehouses now have automated capabilities. The level of automation ranges from a small conveyor belt transporting products in a small area all the way up to a fully automated facility where only a few people are needed to handle storage activity for thousands of pounds/kilograms of product. In many cases, warehouses use machines to handle nearly all physical distribution activities such as moving product-filled pallets (that is, platforms that hold large amounts of product) around buildings that may be several stories tall and the length of two or more football fields.

- **Climate-controlled warehouses:** Warehouses handle storage of many types of products, including those that need special handling conditions such as freezers for storing frozen products, humidity-controlled environments for delicate products, such as produce or flowers, and dirt-free facilities for handling highly sensitive computer products. These are typically used for agricultural products, but there are other items besides food that may require temperature/humidity control, such as some medical products.

Warehouse Strategy

The actual warehouse strategy of an organization may vary by industry, volume, seasonality pattern, and their competitive strategy. The fact is that many firms utilize a combination of private, public, and contract facilities.

For example, a private or contract facility may be used to cover basic year-round requirements, whereas public facilities are used to handle peak seasons. In other cases, central warehouses may be private, whereas market area or field warehouses are public facilities.

A good rule of thumb when planning your warehouse strategy is that a warehouse designed for full-capacity utilization will be fully utilized between 75% and 85% of the time. (That is, 15% to 25% of the time is only used for *surge capacity* to meet peak requirements.) In this case, it may be more efficient to build private facilities to cover the 75% requirement and use public facilities to accommodate peak demand.

In other cases, a firm may find that private warehousing is the best route to go at specific locations on the basis of greater distribution volume, whereas in other markets, public facilities may be the lower-cost option.

In still other cases, the warehouse strategy may be differentiated by customer and product, where some customer groups are best served from a private warehouse, whereas a public warehouse may be the better choice for other customers.

Warehouse Economic Benefits

Warehousing provides specific economic benefits. As discussed earlier, in supply chain and logistics management, there are many cost and service tradeoffs (see Figure 8.1). In general, the more warehouses or DCs you have in your network, the greater the inventory costs but the lower the cost of lost sales (as a result of being closer to your markets).

This is due to the fact that you are trying to hit a *finer* target (that is, smaller demand forecast quantity) and therefore need to carry additional inventory, resulting in higher carrying costs. This is described by a theory known as the *square root rule*, which states that total safety stock can be approximated by multiplying the total inventory by the square root of the number of future warehouse locations divided by the current number.

For example, if you are going from one to two DCs in your network, you would need to carry 40% more safety stock (that is, the square root of 2 / 1, which equals 1.41).

As we carry more inventory, warehouse operations costs tend to increase, but due to the economies we gain by having what is known as a *long in, short out*, our transportation costs decline. The idea here is that, depending on our demand profile, we are able to ship full truckloads to our markets serviced by a DC as opposed to many more-expensive smaller shipments.

More specifically, the benefits can be categorized as described in the following subsections.

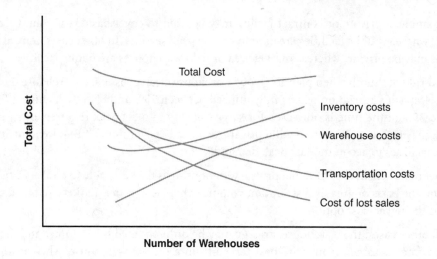

Figure 8.1 Supply chain and logistics management cost tradeoffs

Consolidation

This is a form of warehousing that pulls together small shipments from a number of suppliers in the same geographic area and combines them into larger, more economical shipping loads intended for the same area as described in the *long in, short out* rule mentioned earlier (see Figure 8.2).

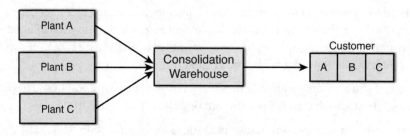

Figure 8.2 Consolidation warehouse

Consolidation warehousing may be used by a single firm, or a number of firms may join together and use a for-hire consolidation service.

Through the use of a consolidation program, each individual manufacturer or shipper can have lower total distribution cost than could be realized on a direct shipment basis individually.

Break-bulk warehouse operations are similar to consolidation except that no storage is performed. This is the term typically used to describe the regional warehouses used by LTL motor carriers.

A break-bulk operation receives combined customer orders from manufacturers and ships them to individual customers and sorts or splits individual orders and arranges for local delivery (see Figure 8.3). As the long-distance transportation movement is a large shipment, transport costs are lower, and there is less difficulty in tracking.

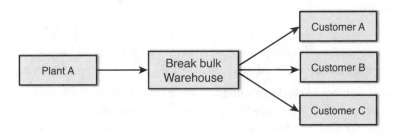

Figure 8.3 Break-bulk warehouse

Accumulation, Mixing, and Sorting

Accumulation may occur when a warehouse accumulates *spot stock* for a selected amount of a firm's product line to fill customer orders during a critical marketing period. Manufacturers with limited or highly seasonal product lines are users of this type of service.

Instead of placing inventories in warehouse facilities on a year-round basis or shipping directly from manufacturing plants, delivery time can be substantially reduced by advanced inventory commitment to strategic markets.

When warehouse facilities are utilized for stock spotting, it allows inventories to be placed in a variety of markets adjacent to important customers just before a peak period of seasonal sales.

This is also used by suppliers of agricultural products to position their products closer to a service-sensitive market during the growing season. After the growing season, the remaining inventory is shipped back to a central warehouse.

Accumulation can also help to provide an assortment to customers and thus provides the benefit of mixing (or sorting), where truckloads of products are shipped from manufacturing plants to warehouses (see Figure 8.4). Each large shipment enjoys the lowest possible transportation rate.

Upon arrival at the mixing warehouse, factory shipments are unloaded, and the desired combination of each product for each customer or market is selected.

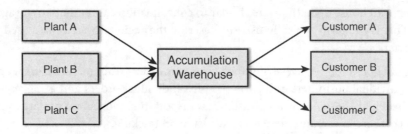

Figure 8.4 Accumulation warehouse

When plants are geographically separated, overall transportation charges and warehouse requirements can be reduced by mixing.

An assortment warehouse stocks product combinations in anticipation of customer orders. The assortments may represent multiple products from different manufacturers or special assortments as specified by customers.

In the first case, for example, a uniform wholesaler would stock products from a number of different clothing suppliers so that customers can be offered assortments.

In the second case, the wholesaler would create a specific company uniform, including shirt, pants, and shoes.

Postponement

Warehouses can also be used to postpone, or delay, production by performing processing, light manufacturing, assembly, and labeling activities. A warehouse with packaging or labeling capability allows postponement of final production until actual demand is known. For example, when I was at Arm & Hammer, we used public warehousing for our finished goods distribution. Many of our larger customers such Walmart and ShopRite wanted to *differentiate* our product, so they required us to put special stickers on the items. We paid the warehouse to unpack and label to order. Although this was not very efficient, it significantly reduced the number of unique stock keeping units (SKUs) that we had to keep in stock, and thus we could label them to order, which lowered our risk and our inventory carrying costs.

Allocation

This involves the matching of on-hand inventory to customer orders in the packaging configuration desired by the customer. For example, a wholesaler may order product from its supplier in pallet quantities, which are made up of many cases of the same product. Their customer, a small retailer, may only want one case or even individual bottles of product contained in a case. The concept of allocation allows for this to occur.

Market Presence

Market presence might not seem to be an obvious benefit of a warehouse; the idea of having a local warehouse is often mentioned by marketing managers as a way to gain a competitive advantage. The market presence factor is based on the perception or belief that local warehouses can be more responsive to customer needs and offer quicker delivery than more distant warehouses. In many cases, especially in the grocery industry, customers have their own fleet of trucks, which they can use to pick up their orders from a supplier instead of having it delivered, thus saving the customer on shipping costs.

As a result, it is thought that a local warehouse will enhance market share and potentially increase profitability.

Warehouse Design and Layout

The first decision is where to locate a warehouse, which can be a strategic decision using a variety of methods, as discussed later in this book.

Once the site is selected, warehouse design is performed using criteria that looks at both physical facility characteristics and product movement.

One of the first facility considerations is to decide the size of the facility within the network.

Size of Facility

One of the major determinants is the demand that is expected to be stored and distributed through the facility now and in the foreseeable future. This is effected by the product mix, functional requirements such as allocation methods (for example, *pallet in, pallet out, case pick*, or *pick and pack* by item), automation, and so on. An area may also be needed for processing rework and returns, office space might be needed for administrative and clerical activities, and space must be planned for miscellaneous requirements and any other value-added functions performed such as kitting, light assembly, and so on.

There are three factors that need to be considered in the design process, as covered in the following subsections.

Number of Stories in the Facility

The ideal warehouse design is limited to a single story so that product does not have to be moved up and down. The use of elevators to move product from one floor to the next requires time and energy.

Conveyors and elevators are also often a bottleneck in product flow because many material handlers are usually competing for a limited number of elevators.

Although it is not always possible, particularly in central business districts where land is restricted or expensive, warehouses should be limited to a single story.

Cube Utilization

The objective is to optimize tradeoffs between handling costs and costs associated with warehouse space. You want to maximize the total *cube* of the warehouse (that is, utilize its full volume while maintaining low material handling costs).

The fact is that warehouse density tends to vary inversely with the number of different items stored, which may seem counterintuitive. The reason is that if you only had one SKU in your warehouse, all the containers would have the same exact cubic dimensions, making it fairly easy to maximize the density of what is stored there, getting the most for your investment. The reality is that in most warehouses, a variety of items in various sizes are stored, making it harder to maximize storage density.

Regardless of facility size, the design should maximize the usage of the available cubic space by allowing for the greatest use of height on each floor.

Older warehouses tend to have 20- to 30-foot ceilings, although modern automated and high-rise facilities may have ceiling heights up to 100 feet.

Through the use of racking or other hardware, it should be possible to store products up to the building's ceiling, which may allow for up to eight racks high.

Maximum effective warehouse height is limited by the safe lifting capabilities of material-handling equipment, such as forklifts.

Product Flow

Warehouse design should also allow as much straight product flow through the facility as is possible, whether items are stored or not. In general, this means that product should be received at one end of the building, stored in the middle, and then shipped from the other end (not always the case). The reasoning for this is that straight line product flow tends to minimize congestion and confusion.

Where you store product in a warehouse (also known as *product slotting*) can have a huge impact on efficiency and can improve labor productivity as follows:

- Locating product in the best pick sequence can reduce order-picking labor requirements.
- Matching product unit loads with the appropriate size storage slot can reduce replenishment labor requirements.
- Balancing workload between operators can reduce response time and improve flow.
- Separating similar products can avoid picking errors and, as a result, increase picking accuracy.

Other benefits of efficient slotting include the following:

- Lower product damage as a result of storing by heavier product first in the pick path, then more easily damaged product later.

- Palletizing productivity can be improved by sorting product by case height. This results in tighter pallets for better trailer utilization.

- Building expansion can be put off as a result of having optimum warehouse layout and cube utilization.

- In the case of retail DCs, store-level productivity can be increased by organizing product in family groups. This reduces sorting of product for restocking at stores.

Facility Layout

As mentioned, the layout of a warehouse should be designed to maximize flow of material, people, equipment, and even information. (Figure 8.5 shows an example of storage plan designed to maximize product flow.)

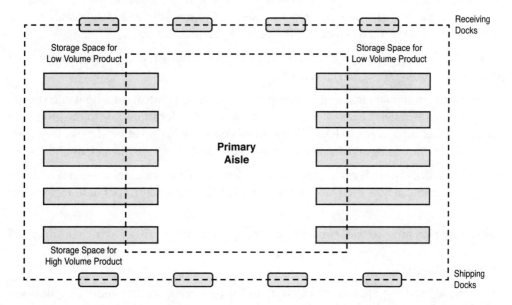

Figure 8.5 Product flow storage plan

It is better for a material handler or piece of handling equipment to make a longer move than to have a number of handlers make numerous, individual, short segments of the same move. Exchanging the product between handlers or moving it from one piece of equipment to another wastes time and increases the potential for damage. In general, fewer, longer movements in the warehouse are preferred.

Where possible, all warehouse activities should handle or move the largest quantities possible. Instead of moving individual cases, warehouse activities should be designed to move groups of cases such as pallets or containers.

This grouping or batching might mean that multiple products or orders must be moved or selected at the same time, sometimes referred to as *wave* or *batch picking* of orders for loading and shipping.

This might increase the complexity of an individual's activities because multiple products or orders must be moved, but it reduces the number of activities and the total cost.

We also need to consider product characteristics, especially those relating to volume, weight, and storage.

Product volume or *velocity* is the major consideration when determining a warehouse storage plan.

High-volume sales or throughput product should be stored in a location that minimizes the distance it is moved, such as near primary aisles and in low storage racks, because such a location minimizes travel distance and the need for extended lifting.

In contrast, lower-volume product can be stored in locations that are farther away from primary aisles or higher up in storage racks.

The storage plan should also have a strategy for products dependent on weight and storage characteristics. Relatively heavy items should be assigned to locations low to the ground to minimize the effort and risk of heavy lifting, often considered as a part of what is called *ergonomics* (that is, designing and arranging things so that people can use them easily and safely).

Bulky or low-density products usually require extensive storage volume, so open floor space or high-level racks may be used for them, whereas smaller items may be put in storage shelves or drawers.

The storage plan must consider and address the specific characteristics of each product.

Material Handling

The material handling equipment necessary will also be based on the product physical and volume characteristics.

The goal of material handling is to increase the cube utilization by using as much height as possible, keep aisle space to a minimum, improve operating efficiency, increase the load per move, and improve speed of response.

Material handling costs are associated with all movement of materials within a warehouse, including incoming transport, storage, finding and moving material, and outgoing transport.

These costs not only include the operators and equipment cost but also supervision, insurance, and depreciation.

The major types of material handling equipment are as follows:

- **Storage and handling equipment:** This type of equipment usually includes non-automated storage equipment including pallet racking and shelving.
- **Engineered systems:** These are custom engineered material-handling systems such as conveyors, handling robots, AS/RS (automated storage and retrieval), AGV (automated guided vehicle), and most other automated material-handling systems.
- **Industrial trucks:** These pieces of equipment are operator-driven motorized warehouse vehicles (for example, forklift truck), powered manually (for example, hand truck) or by gasoline, propane, or electrically.
- **Bulk material handling:** This equipment is used to move and store bulk materials such as liquids and cereals. This equipment is often seen on farms, shipyards, and refineries.

Pallet Positioning

The actual layout of a warehouse will depend on the proposed material handling system and will require development of a floor plan to facilitate product flow (see Figure 8.5).

Warehouse layouts must be refined to fit specific needs, but in general, if pallets will be utilized (as is most common), the first step is to determine the pallet size. While a pallet of non-standard size may be desirable for specialized products, it is best to use standardized pallets because of their lower cost. The most common sizes are 40 by 48 inches and 32 by 40 inches.

In general, the larger the pallet load, the lower the cost of movement per package over a given distance. However, keep in mind that the items to be placed on the pallet and the related patterns will determine, to a large degree, the size of pallet best suited to the operation. No matter the size finally picked, management should adopt one size for the total operation if possible.

It is also important to consider pallet positioning. In a mechanized warehouse, the best and most accessible position is usually a 90-degree, or square, placement, meaning that the pallet is positioned perpendicular to the aisle.

Pilferage and Deterioration

An organization also needs to consider pilferage and deterioration when laying out a warehouse (both the interior and exterior grounds).

Pilferage

It is critical to protect against theft of merchandise in warehouse operations especially for high-value goods.

Typically, at the entrance, as standard procedure, only authorized personnel are permitted into the facility and surrounding grounds, and entry to the warehouse yard should be controlled through a single gate.

Many companies that I've visited have both an external gate and internal security entrance, which operates much like airport security with x-ray wanding of the body, checking bags, and walking through x-ray machines.

Many shortages are the result of innocent mistakes made during order picking and shipment, but the purpose of security is to restrict theft from anywhere. To show how creative theft can be, years ago I worked at an upscale retail DC. At the time, the Izod alligator was extremely popular. It was so valuable that temporary employees were cutting the embroidered alligator logo off of shirts, leaving the shirts behind when they left for the day. So theft, in many cases, can also come from employees during working hours.

It is also important that items are not released from the warehouse unless accompanied by a release document, and if samples are authorized for salespeople, the merchandise should be kept separate from other inventory and also accounted for.

Deterioration

A variety of things can make an item nonusable or nonmarketable (the most common of which is damage from careless transfer or storage). Only so much can be done to protect product from being damaged by rough handling, such as forks accidentally rammed through product by a forklift driver or product falling off of racks.

Product that is noncompatible with other products but stored in the same facility can also cause deterioration (or contamination), but deterioration from careless handling within the warehouse is the most common form and a loss that cannot be insured.

Warehouse Operations

The major processes in a warehouse are receiving, putaway, storage, picking, loading, and shipping (see Figure 8.6). Each involves a series of steps.

- **Receiving:** This process can include the scheduling of carriers to deliver product and the unloading, counting, verifying of order or bill of lading information, and inspecting of material.

  Receiving usually entails a combination of lift truck, conveyors, and manual processes. When product is floor stacked, it may need to be manually offloaded and put onto a pallet or conveyor. Unit or pallet loads can be offloaded with forklift trucks for efficient unloading of a trailer or container.

- **Putaway:** Product is identified by SKU or part number, sorted, and put away, with the location being recorded. Product is usually brought to remote storage.

- **Storage:** Product is stored in a designated location within the warehouse. This may be racks, bins, shelves, or even on the floor.

- **Picking:** This process occurs when customer orders for product are ready to be shipped or to replenish a forward case or unit pick area. In the case of replenishment, pick slots are replenished from storage locations (typically full pallets or cases). Depending on the size of the orders (that is, full pallets, case, or *eaches*), outbound orders are picked from either storage or pick-and-pack areas, SKUs and quantities are then validated, and the items are brought to shipping area.

- **Shipping:** The shipping process includes the scheduling of the carrier for pickup as well as the staging of product in the shipping dock area and finally loading onto a vehicle.

Once the carrier arrives, the order (both product and documents) is prepared for shipment and then loaded and secured on the correct vehicle with protective packaging for shipment and documents. Dispatch is then called, and the product is sent on its way.

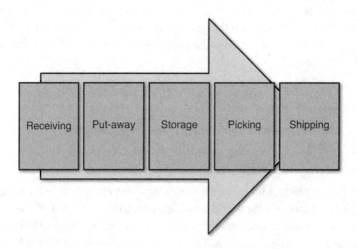

Figure 8.6 Major warehouse processes

Packaging

Everything we buy, whether in business or as a consumer, comes in some kind of packaging. From a logistics perspective, packaging provides protection for the item as it being handled in the warehouse or when the item is being shipped, as well as for economies and efficiency in shipping, storage, and handling.

As mentioned previously, most warehouses use pallets. Therefore, it is important when developing packaging that items are stored safely and efficiently on a pallet, thus reducing the cost of materials handling. Customers who will be purchasing items at the pallet level will also benefit from efficient packaging.

Packaging must protect the item from damage during handling and from environmental damage such as extreme temperature, water damage, contamination with other goods, or damage from static in the case of electronic items.

Internal packaging is developed mainly as a visual device to interest to appeal to the consumer as well as information required by law. The external packaging protects the internal packaging and the item and must also have enough information so that it identifies the contents, in text and often with barcodes for use with radio frequency (RF) technology in the warehouse. The external packaging also needs to have dimensions that allow a reasonable quantity to be stored on a pallet in the most efficient manner.

When companies develop packaging, they usually look at lightweight materials where possible, such as paperboard, aluminum, and plastic.

Corrugated cardboard is typically used for efficient exterior packaging, as a result of its strength, light weight, and recyclability. The corrugated outer container has information printed on it, as well as barcodes that identify the item and manufacturer, and in some cases radio frequency identification (RFID) tags, which use wireless noncontact RF electromagnetic fields to store and transfer data, for the purposes of automatically identifying and tracking tags attached to items.

Key Metrics

In a warehouse, there are both customer-facing and internal metrics.

Customer-Facing Metrics

The customer-facing metrics look at order accuracy and completeness, because customers want to receive the exact products and quantities that they ordered at the right time, not substitute or incorrect items, and/or wrong quantities that are shipped late. Timeliness is a critical component of customer service. There is one measurement that is kind of the white whale in the logistics field because it is very difficult to attain, known as the *perfect order measurement*, which measures whether an order (by line item) is delivered to the right place, at the right time, in defect-free condition and with the correct documentation, pricing, and invoicing.

So, for example, consider an order with the following metrics:

> Order entry accuracy: 99.9% Correct (10 errors per 10,000 order lines)
>
> Warehouse pick accuracy: 99%
>
> Delivered on time: 97%
>
> Shipped without damage: 99%
>
> Invoiced correctly: 99.5%

It has a perfect order measure of 99.9% * 99% * 97% * 99% * 99.5% = 94.4%.

Internal Metrics

Internal measurements look at speed and efficiency or productivity from a variety of views, including distribution cost and aggregate cost efficiency (total distribution spending versus goal or budget), asset utilization, and resource productivity and efficiency because distribution costs can average as much as 10% of every sales dollar.

Technology

Two general types of software are commonly used today at warehouses and DCs. They are *warehouse management systems* (WMSs) and *yard management systems* (YMSs).

Warehouse Management Systems

WMSs are used to manage the receipt, movement, and storage of materials within the four walls of a facility and process the related transactions necessary for receiving, putaway, picking, packing, and shipping. The best-in-class systems go beyond basic picking, packing, and shipping and use advanced algorithms to mathematically organize and optimize warehouse operations.

Many enterprise resource planning (ERP) vendors include WMS modules, but more typically, companies license from vendors that specialize in this type of system, and they are then integrated with their ERP or accounting systems and can be installed systems or cloud-based, on-demand software-as-a-service (SaaS) systems (see the WMS example in Figure 8.7).

Warehouse management systems can use automatic identification and data-capture technology, such as barcode scanners, mobile computers, and potentially RFID, to efficiently manage and monitor the flow of products, because speed and accuracy are paramount in a warehouse. Once data has been collected, data is synchronized either via batch or real-time wireless transmission to a central database, which provides a variety of reports about the status of material in the warehouse.

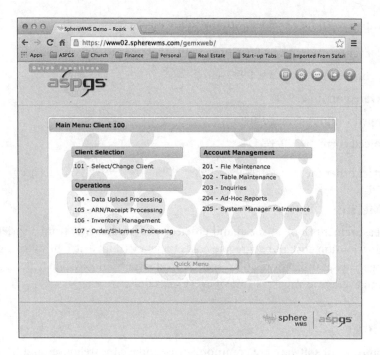

Figure 8.7 Warehouse management screen (Reprinted with permission of SphereWMS from ASP Global Services)

In warehouses where there are multiple picking locations requiring fast and accurate picking, a *pick-to-light* or light-directed system can be used to enhance the capabilities of the employees. A pick-to-light system has lights above the racks or bins the employee will be picking from. The operator then scans a barcode that is on a tote or picking container representing the customer order. Based on the order, the system will require the operator to pick an item from a specific bin. A light above the bin will illuminate, showing the quantity to pick. The operator then selects the item or items for the order. The operator then presses the lighted indicator to confirm the pick. If no further lights are illuminated, the order is complete.

Voice-directed picking systems are gaining popularity. In this type of picking system, workers wear a headset connected to a small wearable computer that tells the worker where to go and what to do using verbal commands. The operators then confirm their tasks by saying predefined commands and reading confirmation codes printed on locations or products throughout the warehouse. These systems are used instead of paper or mobile computer systems requiring workers to read instructions and scan barcodes or enter information manually to confirm their tasks, thus freeing the operator's hands and eyes.

Yard Management Systems

YMSs integrate warehouse operations with inbound and outbound transportation and maximize yard and warehouse efficiency by managing the flow of all inbound and outbound goods.

The YMS enables a business to plan, execute, track, and audit loads based on critical characteristics such as shipment type, load configuration, labor requirements, and dock and warehouse capacity.

The YMS is used to arrange dock appointments for receiving orders and for arranging and scheduling outbound transportation equipment and also helps to manage materials and transportation equipment in the warehouse or factory yard.

Many WMS and most, if not all, ERP systems include customer order processing and management functionality, which we cover next.

9

Order Management and Customer Relationship Management

So far, we have discussed planning for demand and supply and how transportation and warehouse operations provide time and place utilities to the customer. Ultimately, it falls upon the shoulders of the supply chain and logistics function to fulfill orders to meet customer demand, and that is where order management and customer relationship management come into play.

Order management refers to the set of activities that occur between the time a company receives an order from the customer and the time a warehouse is notified to ship the goods to fill that order. Another term, *order fulfillment,* includes the steps involved in receiving, processing, and delivering orders to end customers. In many cases, they are used interchangeably.

The actual time that it takes to perform these activities is often referred to as the *order cycle* or *lead time* (see Figure 9.1), and some organizations expand on this to include customer payment, which is referred to as the *order-to-cash cycle.*

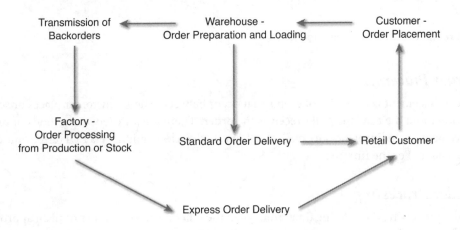

Figure 9.1 Customer order cycle

Wrapped around this process is customer service, which has expanded to be known as *customer relationship management*, or CRM. CRM refers to focusing on customer requirements and delivering products and services, resulting in high levels of customer satisfaction, and also refers to automated transaction and communication applications to support this function.

Order management or fulfillment and customer service are usually part of the supply chain or operations functions because they are intimately related.

Order Management

Order management is primarily made up of four stages: order placement, order processing, order preparation, and loading and order delivery (see Figure 9.2).

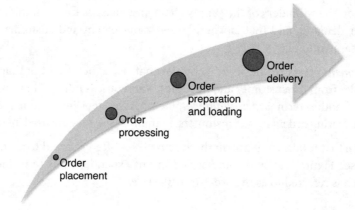

Figure 9.2 Order management process

Order Placement

Order placement is the series of events that occur between when a customer places or sends an order and the time the seller receives the order. There are a variety of methods of order placement, including in person, mail, telephone, fax, or electronically via EDI (electronic data interchange) or the Internet.

Order Processing

Order processing refers to the time from when the seller receives an order until an appropriate location (that is, warehouse) is authorized to fill the order.

Order processing typically may include the following steps:

1. Check for completeness and accuracy.

2. Customer credit check.

3. Order entry into the computer system (manually or electronically).

4. Marketing department credits salesperson.

5. Accounting department records the transaction.

6. Inventory department locates the nearest warehouse to the customer and advises them to pick the order (again, manually or transmitted electronically, depending on the company's technological capabilities).

7. Transportation department arranges for shipment of the order.

Factors that may affect order processing time include the following:

- **Processing priorities:** Similar to short-term scheduling, can be first-come, first-served, shortest lead time, and so on. This varies based on an individual organization's strategy and policies.

- **Order-filling accuracy:** The more accurate the order is received from the customer and input by customer service, the less time spent correcting it.

- **Order batching:** It may be more efficient to batch orders for picking in a warehouse. One batching method, mentioned previously, is known as *wave picking,* where orders are assigned into groupings or waves and released together.

- **Lot sizing:** Full pallet orders may be processed faster than case or unit pick, as discussed in Chapter 8, "Warehouse Management and Operations."

- **Shipment consolidation:** Full truckload orders will be delivered faster than less than truckload (LTL), as discussed Chapter 7, "Transportation Systems." Consolidating small orders going to the same area can not only decrease transportation costs but also speed delivery.

Order Preparation and Loading

Order preparation and loading includes all activities from when an appropriate location is authorized to fill the order until goods are loaded aboard an outbound carrier.

In many cases, this can be one of the best places to improve the effectiveness and efficiency of an order cycle and can account for the majority of a facility's operating cost and time. Technology such as handheld scanners, radio frequency identification (RFID), voice-based order picking, and pick-to-light systems (discussed in Chapter 8) can help speed up the process.

Order Delivery

Order delivery is the time from when a carrier picks up the shipment until it is received by the customer. It is important to closely coordinate picking and staging of orders with carrier arrival because docks and yards can get congested easily and charges apply when carriers are made to wait too long before loading.

Most consumer goods are delivered either from a point of production (factory or farm) in the case of larger or expedited shipments or more typically through one or more points of storage (that is, manufacturer, wholesaler/distributor, or retail warehouses) to a point of sale (that is, retail store), where the consumer buys the good to consume there or to take home.

There are many variations on this model for specific types of goods and modes of sale. Products sold via catalog or the Internet may be delivered directly from the manufacturer or field warehouse to the consumer's home. In some cases, manufacturers may have factory outlets that serve as both a warehouse and a retail store.

While all processes in the supply chain in general, and order management specifically, are subject to measurement (covered later in the text), this is where the "rubber meets the road," so to speak, and is probably one of the most critical points for success or failure in the supply chain. This is due to the fact that delivery, which may be performed by a third party in many cases, is the last point of physical contact with the customer (except in the case of returns, covered in the next chapter).

Customer Relationship Management

The supply chain and logistics function supports a vast array of entities, including suppliers, manufacturers, distributors, wholesalers, and retailers, and so the "customer" can wear many hats.

From the view of the general supply chain, the consumer is the final customer. However, depending on where you are in the supply chain, the customer may be the next step in the supply chain. If you are a supplier or vendor, a manufacturer may be your customer. To the manufacturer, the wholesaler, distributor, retailer, or end user may be your customer. Within each "node" in the supply chain (for example, manufacturing), there are a host of processes, so the next step in the process may be your customer as well as the end user.

Customer Service

In general, customer service is a means by which companies try to differentiate their product, sustain customer loyalty, increase sales, and improve profitability. Its main elements are price, product quality, and service.

The supply chain and logistics function is a critical part of the marketing mix because it has a significant impact on all four of its components of product, price, promotion, and place.

As discussed earlier this book, the supply chain provides a place and time utility to the customer as provided by the logistics or physical distribution variables of product availability and order cycle time.

Multifunctional Dimensions of Customer Service

Looking at customer service in terms of its multifunctional dimensions, logistics can provide benefits to the customer in terms of the following:

- **Time:** Meaning the entire order fulfillment cycle time
- **Dependability:** Can offer guaranteed fixed delivery times of accurate, undamaged orders
- **Communications:** Offers ease of order taking, and queries response
- **Flexibility:** Provides the ability to recognize and respond to a customer's changing needs

Transactional Elements of Customer Service

Customer service can be looked at as having three transactional components (see Figure 9.3):

- **Pre-transaction elements:** Customer service factors that arise prior to the actual transaction taking place.
- **Transaction elements:** The elements directly related to the physical transaction and are those that are most commonly concerned with logistics.
- **Post-transaction elements:** These involve those elements that occur after the delivery has taken place.

The most dominant customer service elements are logistical in nature, and late delivery is the most common service complaint, with speed of delivery usually being one of the most important service elements. The penalty for service failure is primarily lost sales.

At the end of the day, the most important elements from the customer perspective are usually on-time delivery, order fill rate, product condition, and accurate documentation. As the saying goes, it is a lot more expensive to acquire a new customer than to keep an existing one.

Service Quality and Metrics

Service "quality" is a measure of the extent to which the customer is experiencing the level of service that they are expecting. In other words, service quality is the match between what the customer expects and what the customer experiences. Mathematically, service quality can be expressed as follows: *(perceived performance / Desired expectations) * 100*

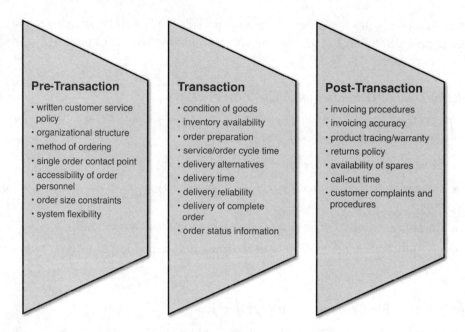

Pre-Transaction
- written customer service policy
- organizational structure
- method of ordering
- single order contact point
- accessibility of order personnel
- order size constraints
- system flexibility

Transaction
- condition of goods
- inventory availability
- order preparation
- service/order cycle time
- delivery alternatives
- delivery time
- delivery reliability
- delivery of complete order
- order status information

Post-Transaction
- invoicing procedures
- invoicing accuracy
- product tracing/warranty
- returns policy
- availability of spares
- call-out time
- customer complaints and procedures

Figure 9.3 Components of customer service

In general, when establishing customer service objectives, they should be specific, measurable, achievable, and cost-effective.

Chapter 4, "Inventory Planning and Control," discussed how safety stock levels can be set to desired service levels, which illustrates the cost/service tradeoff, at least from a product availability standpoint.

Other customer service metrics include the following:

- Percent of sales on backorder
- Number of stock outs
- Percent of on-time deliveries
- Number of inaccurate orders
- Order cycle time
- Fill rate as a percentage of demand met, orders filled complete, and so on

Internal Versus External Metrics

Customer service can also be looked at from an internal (for example, item and line fill rate) and external metric (order fill rate and "perfect order") perspective and should be tied to organizational and functional strategies and can be benchmarked against other best-in-class companies in your industry as well.

Levels of Focus

Another way that some companies look at customer satisfaction is by breaking it into levels of focus.

The first level is a basic focus on customer "service," where you offer a product/service and a customer needs that product/service. The transaction happens, money changes hands, and no major issues come up. You then benchmark it against industry and competitor practices to achieve internal standards.

The second level, or customer "engagement," is about building a relationship and loyalty so that when a customer is ready to buy, they will immediately purchase from the company they've been engaged with during that time. You also want to consider the customer's perception of satisfaction and manage performance to keep them satisfied.

The third level, customer "intimacy," occurs when companies are close enough to their customers that they can begin to anticipate a customer need and respond accordingly. You may even extend the supply chain to include your customer's customer and also provide value-added services for select customers.

Managing Customer Service

To reach the third level of customer "intimacy," it can be useful for an organization to segment individual customers or groups of customers based on profitability. This is known as *customer profitability analysis* (CPA). CPA is the allocation of revenues and costs to customer segments or individual customers to calculate the profitability of the segments or customers (see Figure 9.4).

Product Service	Customer Segment A	Customer Segment B	Customer Segment C
Service Level (on Time)	99%	95%	90%
Order Fill (complete)	99%	98%	95%
Order Lead Time	1 day	3 days	5 days
Delivery Time	24 hours	48 hours	72 hours

Figure 9.4 Customer profitability analysis

CPA suggests that different customers consume differing amounts and types of resources and recognizes that all customers are not the same and that some customers are more valuable than others to an organization.

It can be used to identify when an organization should pursue different logistical approaches for different customer groups and has been facilitated by the acceptance of activity-based costing, which measures the cost and performance of activities, resources, and cost objects.

Resources are assigned to activities, and then activities are assigned to cost objects based on their use. This is as opposed to traditional cost accounting, which is well suited to situations where an output and an allocation process are highly correlated, which might not be the case in the area of supply chain and logistics customer service.

Service Failure and Recovery

There will be situations where actual performance does not meet the customer's expected performance, commonly referred to as a *service failure*. Examples of order-related service failures include the following:

- Lost delivery
- Late delivery
- Early delivery
- Damaged delivery
- Incorrect delivery quantity

The process for returning a customer to a state of satisfaction after a service or product has failed to live up to expectations, known as *service recovery*, can be costly, but it may lead to an increase in customer loyalty. However, unsatisfactory service recovery can increase the consequences of the initial failure.

Customer Relationship Management

Customer relationship management, or CRM, is the act of strategically positioning customers to improve the profitability of the organization and enhance its relationships with its customer base. Tools such as CPA and activity-based costing (ABC; which identifies activities in an organization and assigns the cost of each activity with resources to all products and services according to the actual usage by each) can be used to segment an organization's customer base by profitability and identify the product/service package for each customer segment per Figure 9.4. To be successful, you must then develop and execute the best processes for each segment and measure performance and continuously strive for improvement.

For a CRM strategy to succeed, there must also be an infrastructure that increases value to the customer and a means to motivate them to remain loyal. To improve profitability, the CRM system must manage both relationships among people within an organization and between customers and the company's customer service representatives.

Besides segmenting customers for supply chain customer service, organizations use CRM to target their marketing efforts; use relationship marketing or permission marketing, where customers select the type and time of communication; and cross-selling (as well as upselling), where additional products are sold as the result of an initial purchase (for example, emails from barnesandnoble.com describing other books bought by people that match your tastes).

Technology

An order management system (OMS) is a computer software system used in many industries for order entry and processing. In most cases, it is part of a larger ERP or accounting system (see Figure 9.5).

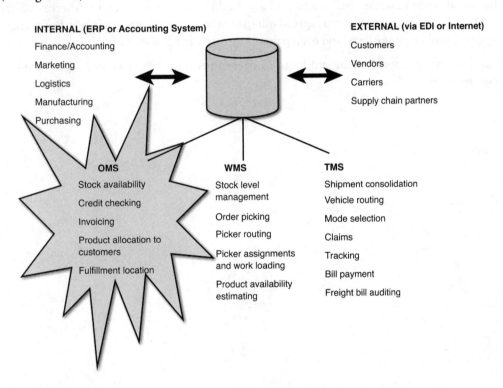

Figure 9.5 Order management system (OMS) and other supply chain execution systems

OMS applications manage processes, including order entry, customer credit validation, pricing, promotions, inventory allocation, invoice generation, sales commissions, and sales history.

A distributed order management (DOM) system differs from an OMS in that it manages the assignment of orders across a network of multiple production, distribution, and retail locations to ensure that logistics costs and customer service levels are optimized.

An OMS is usually deployed as part of an enterprise application, such as an ERP system, because its sales engine is integrated with the organization's inventory, procurement, and financial systems.

As previously mentioned, a CRM system manages a company's interactions with current and future customers and involves using technology to organize, automate, and synchronize sales, marketing, customer service, and technical support. This includes the management of business contacts, clients, contract wins, and sales leads within the sales function, sometimes referred to as *sales force automation* (SFA) software.

Perhaps the biggest benefit to most businesses when moving to a CRM system comes from having all your business data stored and accessed from a single location, whereas before CRM systems, customer data was spread out over office productivity suite documents, email systems, mobile phone data, and even paper note cards and Rolodex entries.

The last step in supply chain and logistics operations is known as *reverse logistics* and is covered next.

10

Reverse Logistics and Sustainability

Reverse logistics, an often-overlooked process that can help companies reduce waste and improve profits, is, as the name implies, the reverse of what we've described so far in terms of planning and operations. It can be defined as the process of planning, implementing, and controlling the efficient flow of recyclable and reusable materials, returns, and reworks from the point of consumption for the purpose of repair, remanufacturing, redistribution, or disposal

To take this a step further, with today's environmental concerns, organizations need to try to integrate environment thinking into the entire supply chain process, forward and reverse. This includes product design, material sourcing and selection, manufacturing processes, delivery of the final product to the consumers, and end-of-life management of the product after its useful life.

In a perfect world, of course, there would be no need for much of the material handled in reverse logistics, and later in the book, we discuss ways to use Lean thinking to reduce waste in the supply chain, but for now we will examine concepts and applications in this area.

Reverse Logistics Activities

According to a 2010 Aberdeen Group study, the average manufacturer spends an astounding 9% to 15% of total revenue on returns (Aberdeen Group, 2010).

There are a variety of reasons for the reverse logistics process, including the following:

- Processing returned merchandise, including damaged, seasonal, restock, salvage, recall, or excess inventory
- Green initiatives such as recycling packaging materials/containers
- Reconditioning, refurbishing, remanufacturing of returned product
- Disposition of obsolete inventory
- Hazardous materials recovery and electronic waste disposal

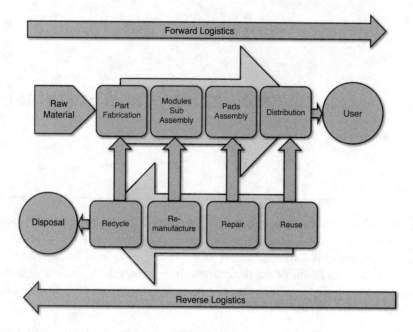

Figure 10.1 Logistics and reverse logistics processes

So, depending on the specific reasons for the process existing in the first place within an organization, the reverse logistics network can be used for a variety of purposes, such as refilling, repairs, refurbishing, remanufacturing, and so on, depending on the nature of the product, unit value, sales volume, and distribution channels.

Let's look at some of the major reasons mentioned here in some more detail.

Repairs and Refurbishing

Repair is a regular feature in service-based products under a warranty period, and almost all consumer durables need repairs on a regular basis. Refurbishing, in contrast, is applied to goods returned because of damage, defects, or below-promised performance during the warranty period.

Manufacturers may establish the reverse logistics system not only for offering free service during the warranty period but also for extending the services beyond the warranty period on a chargeable basis.

The system usually operates through the company's service centers where repair and refurbishing takes place.

The physical collection of defective products is performed through a dealer network. The collected products are sent to the nearest service center for overhaul, repairs, or refurbishing.

Refilling

Reverse logistics is integrated to an organization's supply chain in the cases of the reusable nature of packages such as glass bottles, plastic containers, print cartridges, and so on.

In case of large refillable water bottles, for example, the delivery truck delivers filled bottles to and collects the same number of empty bottles from them for delivery to the factory. No extra transportation costs are involved in the process because the same delivery truck originates and terminates its journey at the factory where these reusable bottles are refilled for redelivery to customers.

Typically, this type of arrangement is accomplished via a *hub-and-spoke* distribution system (that is, a centralized distribution system where inventory is shipped from a central location to smaller locations or directly to consumers, similar to a bicycle hub-and-spoke configuration).

Recall

This is an emergency situation where the products distributed in the market are called back to the factory because of any of the following reasons:

- Product not giving the guaranteed performance
- Quality complaints from many customers
- Defective products causing harm to human life
- Products beyond expiry date
- Products with defective design
- Incomplete product
- Violation of government regulations
- Ethical considerations
- Save the company image

A product recall puts a large financial burden on a company, but in the competitive scenario, the companies consider *recall* as an opportunity to increase customer satisfaction.

Remanufacturing

Manufacturers in developed countries are putting in practice a relatively new concept of remanufacturing because during the usage of the product, it undergoes wear and tear. During remanufacturing, worn-out parts are replaced with new ones, and the performance of the product is upgraded to the level of a new one.

Similarly, equipment sold can be checked after use to the remanufacturing process and be brought back to the remanufacturing unit.

The investment in remanufacturing and related reverse logistics supply chain can be justified on the basis of economies of scale.

Recycling and Waste Disposal

Leftover materials, used product, and package waste are causing environmental pollution and creating problems for disposal.

In many countries, governments are devising regulations to make manufacturers responsible to minimize waste by recycling products.

Returns Vary by Industry

In some industries, returns are the major reason for a reverse logistics system; percentages can range from as low as 2% to 3% (chemicals) to more than 50% (magazine publishing).

Let's look at some industries to examine why the return rates are so high.

Publishing Industry

The publishing industry has the highest rate of unsold copies (28% on average). This has been partially a result of the growth of large chain stores requiring more books and magazines. To secure a prominent display in superstores, publishers must supply large numbers of books. The fact is that superstores sell less than 70% of books they order, and they have a relatively short shelf life.

Computer Industry

Computers have a relatively short lifecycle, so there are opportunities to reuse and create value out of computer equipment. They contain what is known as *e-waste*, such as lead, copper, aluminum gold, plastics, and glass. E-waste not only comes from computers but also televisions, cell phones, audio equipment, and batteries.

For example, in the remanufacturing of toner cartridges, there are 12,000 remanufacturers, employing 42,000 workers, that sell nearly $1 billion worth of remanufactured cartridges annually.

Automotive Industry

There are three primary areas for reverse logistics in the automotive industry:

- Components in working order are sold as is (for example, parts from junkyards).
- Components such as engines, alternators, starters, and transmissions are refurbished before they can be sold.
- Materials are reclaimed through crushing or shredding.

Automotive recyclers handle more than 37% of the nation's metal scrap, and the remanufactured auto parts market is estimated at $34 billion annually.

Retail Industry

Profit margins in retail are so slim that good return management is critical because returns reduce the profitability of retailers marginally more than manufacturers. In fact, returns reduce the profitability of retailers by 4.3% (Rogers & Tibben-Lembke, 1998).

Reverse Logistic Costs

Reverse logistics costs come from a variety of activities, such as merchandise credits to the customers, transportation costs of moving the items from the retail stores to the central returns distribution center, repackaging of the serviceable items for resale, the cost of warehousing the items awaiting disposition, and the cost of disposing of items that are unserviceable, damaged, or obsolete.

Besides those tangible costs, there are intangibles that greatly impact the customer, such as increased customer wait times, loss of confidence in the supply system, and the placement of multiple orders for the same items.

Reverse Logistics Process

There are five steps to the product returns process, no matter what industry you are in: receive, sort and stage, process, analyze, and support (Stock, Speh, & Shear, 2006).

Receive

Product returns are received at a centralized location, usually a warehouse or distribution center (usually after being gathered from retail locations or returned by the end user themselves). In many cases, a first step in this process is to provide a return acknowledgment.

The returns may arrive via many carriers and in a variety of packages, either on full pallets or individual containers.

The concept of *pre-postponement* can be useful in this process, where companies such as Sauder Woodworking Company, which makes ready-to-assemble furniture, processes returns as close as possible to the point of sale so as to determine quickly which returns were recoverable and which were not.

Sort and Stage

In this stage, returned products are received and sorted for further staging in the returns process.

The sorting can be based on how the items have been returned (that is, pallets, cartons, packets) or the type, size, or number of the return. This process generally takes 3 days or less to accomplish.

Process

Returned products are then subsorted into items, based on their stock keeping unit (SKU) number. They can then be returned to inventory. If they are vendor returns, they are sorted by vendor.

There is usually some kind of processing station where they are processed by order of their receipt, type of product, customer type or location, physical size of the items, and so forth.

Paperwork that came with the return is separated from the item and compared with the electronic records to identify any discrepancies.

Analyze

The value of the returned item is determined by trained employees to see whether it should be repaired or refurbished and which are allowable versus nonallowable returns, for example.

The last part of this step is the marketing of products that have been repackaged, repaired, refurbished, or remanufactured, which are usually shipped to secondary markets.

Support

At this point, returns in good condition such as back-to-stock or -store items are returned to inventory. If the items require repair, refurbishment, or repackaging, then diagnostics, repairs, and assembly/disassembly operations are performed as needed.

Reverse Logistics as a Strategy

As opposed to looking at reverse logistics as a cost center to be minimized (because the reality is that it is only around 4% of total supply chain costs), some forward-looking companies have started looking at it as a strategic weapon to positively impact revenue.

Using Reverse Logistics to Positively Impact Revenue

A recent UPS white paper on reverse logistics found some key areas where companies can positively impact revenue with reverse logistics activities. They are as follows (Greve & Davis, 2012):

- **Returns-to-revenue:** Companies that ensure timely delivery and processing of returns position themselves to save more or earn more from the returned product. From refurbishing, repackaging, and reselling to parts reclamation and recycling, returned products are often untapped sources for revenue. With the secondary, discount market for products continuing to grow, there is even more reasons to think about returns as revenue opportunities.

- **Protecting profits:** Handling returns properly and tracking all activities is critical to help companies avoid fines and penalties from various government regulatory agencies such as the FDA, the Consumer Product Safety Commission, and other state and federal agencies.

- **Customer loyalty:** According to a nationwide survey conducted in 2005, 95% of customers will not buy from a company if they have a bad returns experience. This, in part, explains why companies considered best in class in reverse logistics enjoy a 12% advantage in overall customer satisfaction over their competition

- **Disposal benefits:** Knowing what is returned and where it ends up makes it easier for companies to deal with regulatory issues and evaluate returned stock for possible secondary sales channels.

  There are also other beneficial byproducts to disposing of products, such as avoiding excess inventory carrying costs, avoiding excess taxes and insurance, and managing staff levels.

- **Maximize recovery rates:** Mishandled or completely misplaced returns affect the efficiency of any reverse logistics process, but it also means that products could end up a being a total loss for a company instead of an opportunity for resell or a spare parts resource.

Other Strategic Uses of Reverse Logistics

Reverse logistics can also be strategically used to reduce the risk from buying products that may not be hot-selling items, by adjusting return rates based on popularity of an item.

It can also be used to increase the switching costs of changing suppliers to lock customers in by taking back unsold or defective merchandise quickly, while crediting the customer without delay. Many retailers and manufacturers have liberalized their return policies in recent years due to competitive pressures. One e-tailer, Zappos.com, even encourages returns as a way to increase customer loyalty.

Many companies use reverse logistics to clean out customer inventories, so that they can purchase more new goods. That way, fresher inventories can demand better prices, which in turn protects margin.

Still others use reverse logistics as a form of good corporate citizenship, using the process for altruistic reasons, such as philanthropy. These activities enhance the value of the brand and are a marketing incentive to purchase their products.

There is also the opportunity to recapture value and recover assets, as in some cases a large portion of bottom-line profits is derived from asset recovery programs. The profit comes from materials that were previously discarded.

Legal disposal issues can create a concern because landfill fees are increasing and the options for the disposal of hazardous material are decreasing. So, legally disposing of nonsalvageable materials becomes more difficult (and may be subject to fines if not done properly) (Rogers & Tibben-Lembke, 1998).

Reverse Logistics System Design

The success of reverse logistics system in achieving the desired objectives depends on the efficiency and effectiveness of a number of subsystems, as covered in the following subsections.

Product Location

The first step in the callback process is to identify the product location in the physical distribution system of the firm. Product location becomes more difficult after it is sold and handed over to the customer.

It is a bit easier in the case of industrial or high-value products because of the limited number of customers and personal interaction with the clients due to direct selling.

Product Collection System

Once the product location is identified, the collection mechanism gets into operation.

This can be done either through company's field force, channel members, or third party. However, proper instructions have to be given to motivate the customer for returning the products.

Third parties are often used if it is not an area of expertise for a company and the third party can do it cheaper and more efficiently.

For example, Best Buy (a major electronics retailer) created a business unit... to focus on developing sales of consumer electronics into the secondary markets... No longer was the handling of customer returns, return to vendor, and overstock a cost center sitting in a dark corner, but now it was transformed into a profit center... at Best Buy, maximizing profit in the reverse logistics business is involving partnerships with both new and existing customers as well as manufacturers and third-party service providers (3PSs).

- **Online stores and auctions:** With product testing, inventory management, listing, payment collection, and order fulfillment, Best Buy has built an integrated supply chain to take returned product from the stores and resell it to its value-seeking customers through eBay, a private online store, and other online channels. Best Buy recently acquired Dealtree, its provider of these services.

Trade-in, an online program was launched in 2007, offering customers a fast and easy alternative to selling online themselves, or just letting working products sit in a drawer. It allows customers to recapture economic value, and through the Dealtree technology, Best Buy has instant access to current market value of products.

- **Refurbishment:** Working with a variety of 3PSs, Best Buy has been integrating refurbished products into its warranty replacement program and selling direct to consumers.
- **Recycling:** In 2008, Best Buy began testing offering free recycling in several markets. They have been building up a network of local certified recyclers and plan to roll nationwide in 2009 (*Reverse Logistics Magazine*, 2009).

Recycling or Disposal Centers

These may be the company's plants and warehouses or some fixed location in the reverse logistics network.

The called-back products are inspected before they are further processed for further repairs, refurbishing, remanufacturing, or waste disposal.

Investments in facilities for these activities depend on the objectives of the system, cost implication, complexity of the operations, and expected gains.

Documentation System

Tracing the product location becomes easier if proper documentation is maintained at each channel level. However, at the time of handing over the product to the customer, the detailed information, if collected through proper documentation, can form a good database that can be used in case of product callbacks.

Reverse Logistics Challenges

There are many challenges to running and maximizing the efficiency of the reverse logistics process, including those discussed in the following subsections.

Retailer-Manufacturer Conflict

Inefficiencies in a reverse logistics process can lengthen the time for processing returns, such as the condition and value of the item and the timeliness of response. The buyer and seller have to develop a good working partnership to derive mutual benefit.

Problem Returns and Their Symptoms

Unprocessed returns are easy to observe but some of these other symptoms are not as easily observed, such as the following:

- Returns arriving faster than processing or disposal.
- Large amount of returns inventory held in the warehouse.
- Unidentified or unauthorized returns.
- Lengthy processing cycle times.
- Unknown total cost of the returns process.
- Customers have lost confidence in the repair activity.

Lack of information about the reverse logistics process can result in the process being out of control. As the saying goes, "If you aren't measuring it, you can't manage it."

Cause and Effect

Poor data collection can lead to uncertainty about return causes. By improving the return process, it is possible to decreases costs. It is important to be able to see defective product by problem code in an information system, thus making it possible to track return issues.

Reactive Response

In recent years, government regulation or pressure from environmental agencies has forced companies to begin to focus on an area that is not one of their core competencies. It has not been possible to justify a large investment in improving reverse logistics systems and capabilities. Some have been able to see it as a win-win game by developing strategies mentioned previously such as good corporate citizenship and recapturing value and recovering assets.

Overall, in many companies, management inattention and the lack of importance of reverse logistics, especially handling returns and nonsalable items, has resulted in restrictive policies in this regard. This may be in part due to not wanting their returns being used to cannibalize existing sales. Recently, there seems to be a trend toward reducing or eliminating restrictive policies and attempting to handle returns more effectively to recover value from what can be a valuable resource (Rogers & Tibben-Lembke, 1998).

Managing Reverse Logistics

A research team at the Reverse Logistics Executive Council identified key reverse logistics management elements and examined the return flow of product from a retailer back through the supply chain toward its original source or to some other disposition (see Figure 10.2) (Rogers & Tibben-Lembke, 1998).

These elements, depending on how they are handled, can either positively or negatively impact a company's profitability. The elements are as discussed in the following subsections.

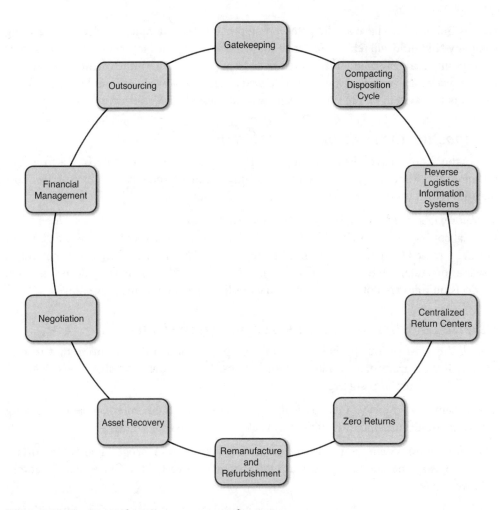

Figure 10.2 Key reverse logistics management elements

Gatekeeping

Gatekeeping is the screening of defective or unwarranted returned merchandise at the beginning of the reverse logistics process.

It is the first critical factor to ensure that the entire reverse flow is both manageable and profitable. In the past, companies have put most resources into the forward logistics process and have given very little time and effort into the reverse process.

While liberal return policies, like those at L.L. Bean, Walmart, and Target, may draw customers, they can also encourage customer abuse, such as the return of items lightly used for an event or one occasion.

So, it is important to have a solid gatekeeping process. For example, the electronic gaming company Nintendo will rebate retailers if they register the game player sold to the consumer at the point of sale. By doing this, Nintendo and retailers can determine whether the product is under warranty, and also if it is being returned inside the allowed time window. The impact from this new system on their bottom line was substantial: an 80% drop in return rates.

Compacting the Distribution Cycle Time

One of the major goals of the reverse logistics process once an item has entered it is to reduce the amount of time to figure out what to do with returned products once they arrive. This includes return product decisions, movement, and processing.

Therefore, it is important to know beforehand what to do with returned goods. Often when material comes back in to a distribution center, it is not clear whether the items are defective or can be reused or refurbished or need to be sent to a landfill. The challenge of running a distribution system in reverse is difficult: Employees have difficulty making decisions when the decision rules are not clearly stated and exceptions are often made.

Reverse Logistics Information Technology Systems

One of the most serious problems that the companies face in the execution of a reverse logistics is the scarcity of good information systems. To work well, a flexible reverse logistics information system is required.

The system should create a database at store level so that the retailer can begin tracking returned product and follow it all the way back through the supply chain.

The information system should also include detailed information programs about important reverse logistics measurements, such as returns rates, recovery rates, and returns inventory turnover.

Useful tools such as radio frequency (RF) are helpful. New innovations such as two-dimensional barcode and radio frequency identification (RFID) license plates may soon be in use extensively.

Centralized Return Centers

Having centralized return centers (CRCs) can offer many benefits to an organization, including the following:

- Consistency in disposition decisions and minimizes errors.
- A space-saving advantage for retailers who want to dedicate as much of the shop floor to salable merchandise as possible.
- Labor cost reductions due to their specialization, because CRC employees can typically handle returns more efficiently than retail clerks can.
- Transportation cost reductions because empty truckloads returning from store deliveries are used to pick up return merchandise.
- A convenient selling tool for the easy disposition of returned items. This can be an appealing service to retailers, and may be a deal maker for obtaining or retaining customers.
- Faster disposition times allow the company to obtain higher credits and refunds, because items stay idle for smaller periods of time, thus losing less value.
- Easier to identify trends in returns, which is an advantage to manufacturer, who can detect and fix quality problems sooner than if these returns were handled entirely by customer service personnel.

Zero Returns

A company may have a program that does not accept returns from its customers. Rather, it gives the retailer an allowable return rate and proposes guidelines as to the proper disposition of the items. Such policies are usually accompanied by discounts for the retailer.

This type of policy passes the returns responsibility onto the retailer, while reducing costs for the manufacturer or distributor.

The drawback is that the manufacturer loses some control over its merchandise.

Remanufacture and Refurbishment

The advantage of remanufacturing and refurbishment is using reworked parts, resulting in a cost savings.

There are five categories of remanufacture and refurbishment:

Make the product reusable for its intended purpose:

1. Repair
2. Refurbishing
3. Remanufacturing

Retrieve reusable parts from old or broken products:

4. Cannibalization

Reuse parts of products for different purpose:

5. Recycling

Asset Recovery

Asset recovery is the classification and disposition of returned goods, surplus, obsolete, scrap, waste and excess material products, and other assets. It tries to maximize returns to the owner, while minimizing costs and liabilities for the dispositions.

The objective of asset recovery is to recover as much of the economic (and ecological) value as is possible, thus reducing the final quantities of waste.

This can be a good cash-generating opportunity for companies, who can sell these goods that would otherwise end up in landfills.

Negotiation

Negotiation is a key element for all parties of the reverse logistics process. Because of the inherent lack of expertise on product returns, negotiations usually are informal and approached without formal pricing guidelines. Firms often do not maximize the residual value of returned product.

Financial Management

This is one of the most difficult parts of reverse logistic and also one of the most important.

Returns are sometimes charged against sales. Sales department personnel may tend to fight returns and delay them as much as possible. Accounts receivables are also impacted by returns.

Outsourcing

As mentioned previously, reverse logistics is usually not a core competence of the firm. In many cases, it may make more sense for the firm to outsource their reverse logistics functions than keep those in-house.

Reverse Logistics and the Environment

In the past, most companies were concerned primarily with the forward logistics processes and to some degree as it relates to returning product to their suppliers. Today, many companies also have a focus on reverse logistics issues because of environmental concerns and, as mentioned before, how it can both add value to the customer and to the bottom line.

Now and into the future, environmental considerations will have a greater impact on many logistics decisions.

Supply Chain Sustainability

As a result of this shifting focus, the term *supply chain sustainability* has become fairly common and refers to the management of environmental, social and economic impacts, and good governance practices throughout the lifecycles of goods and services.

The objective of supply chain sustainability is to create, protect, and grow long-term environmental, social, and economic value for all stakeholders involved in bringing products and services to market.

Green Logistics

Another term has emerged as a result called *green logistics*, which refers to minimizing the ecological impact of logistics. An example of this is a reduction in the energy usage of logistics activities and reduction in the usage of materials. Reducing the carbon footprint in a supply chain is a sustainability priority for logistics.

Environmental considerations have a greater impact on many logistics decisions. For example, many products can no longer be placed in landfills, and as a result, many companies must take back their products at the end of their useful lifetime. At the same time, there is a decrease of landfill availability resulting in an increase in landfill costs.

Many products are banned from being placed in a landfill either because they present a health risk, such as cathode ray tubes (CRTs) in old TVs and computer monitors, or because they take up too much space.

Products that are banned from landfills include the following: motor oil, household batteries, household appliances, paper products, tires, and some medical and electrical equipment. Product bans represent a new reverse logistics opportunity, because when companies are forced to take their products back when they are banned, they reuse the products and recapture their value. The firm is also looked upon as an environmentally friendly company.

Many companies, such as Hewlett-Packard and Xerox, have adopted an extended product responsibility (EPR) program, which focuses on the total life of the product, looking for ways to prevent pollution and reduce resource and energy usage through the product's lifecycle.

Programs and processes like *product takeback* and EPR are part of a strategy that has become known as *closed-loop supply chains*, which are designed and managed to encompass both forward and reverse flows activities in a supply chain.

The reverse logistics activities of reuse, remanufacturing, refurbishing, and recycling have become to be known as the *four R's* of sustainability. The R's, while different, are now being used by many organizations together in a broad program where they complement to each other.

Other examples of companies using green concepts in supply chain to their advantage include the following:

- Walmart, which anticipates its goal of a 5% reduction in packaging by 2013, will produce $3.4 billion in direct savings and roughly $11 billion in savings across the supply chain.

- Johnson & Johnson's energy-efficiency program resulted in an estimated $30 million in annualized savings over the 10 years prior to the company's 2006 sustainability report.

- Nestlé, through a combination of packaging source reduction, reuse, recycling, and energy recovery, saved $510 million, worldwide, between 1991 and 2006 (Futin, 2010).

The emergence of global supply chains has presented challenges, risks, and opportunities for both forward and reverse flows, including environmental or green laws, which is the topic of our next chapter.

11

Global Supply Chain Operations and Risk Management

In the late 1980s, a considerable number of companies began to integrate global sources into their core business, establishing global systems of supplier relationships and expansion of their supply chains across national boundaries and into other continents around the globe.

The globalization of supply chain management in organizations had the goals of increasing their competitive advantage, adding value to the customer, and reducing costs through global sourcing.

In addition to sourcing globally, many companies sell globally or compete with other companies that do.

Ultimately, global supply chain management is about sourcing, manufacturing, transporting, and distributing products outside of your native country. It ensures that customers get products and services that they need and want faster, better, and more cost-effectively either locally or from around the world.

Therefore, we can define global supply chains as worldwide networks of suppliers, manufacturers, warehouses, distribution centers, and retailers through which raw materials are acquired, transformed, and delivered to customers.

Growth of Globalization

In recent years, we have seen a change in how firms organize their production into global supply chains, with companies increasingly outsourcing some of their activities to third parties and locating parts of their supply chain outside their home country (known as *offshoring*).

They are also increasingly partnering with other firms through strategic alliances and joint ventures, enabling not only large but also smaller firms and suppliers to become global.

These types of global business strategies have allowed firms to specialize on *core* competencies to sustain their competitive advantage.

This is not limited to just outsourcing manufacturing and supply chain operations but also includes business process outsourcing (BPO) and information technology (IT) services that are supplied from a large number of locations as well as other knowledge-intensive activities such as research and development (R&D).

Factors Influencing Globalization

Some key factors influence the growth of globalization, including the following:

- **Improvements in transportation:** Larger container ships mean that the cost of transporting goods between countries has decreased. Economies of scale are found as the cost per item can reduce when operating on a larger scale. Transportation improvements also mean that both goods and people can travel more quickly.

- **Freedom of trade:** There are a number of organizations like the World Trade Organization (WTO) that promote free trade between countries, helping to remove barriers between countries.

- **Improvements of communications:** The Internet and mobile technology has allowed greater communication between people in different countries.

- **Labor availability and skills:** Less-developed nations in Asia and elsewhere have lower labor costs and, in some cases, also high skill levels. Labor-intensive industries such as clothing can take advantage of cheaper labor costs and reduced legal restrictions in these less-developed countries.

- **Transnational corporations:** Globalization has resulted in many businesses setting up or buying operations in other countries. When a foreign company invests in a country, by building a factory or a shop, this is sometimes called *inward investment*. Companies that operate in several countries are often referred to as *multinational corporations* (MNCs) or *transnational corporations* (TNCs). The U.S. fast-food chain McDonald's is a large MNC, having nearly 30,000 restaurants in 119 countries.

Many multinational corporations not only invest in other economically developed countries but also invest in less-developed countries as well. (For example, Ford Motor Company makes a large numbers of cars in the United Kingdom and in India.)

Reasons for a Company to Globalize

The reasons a company may choose to globalize vary, but are usually influenced by global, technological, cost, political, and economic influences. Reasons to globalize within each of these influences include the following:

- **Global market forces**

  Foreign competition in local markets

  Growth in foreign demand

  Global presence as a defensive tool

  Companies forced to develop and enhance leading-edge technologies and products

- **Technological forces**

  Knowledge diffusion across national boundaries, hence the need for technology sharing to be competitive

  Global location of R&D facilities

  Close to production (as product cycles get shorter)

  Close to expertise (for example, Indian programmers)

- **Global cost factors**

  Availability of skilled or unskilled labor at lower cost

  Integrated supplier infrastructure (as suppliers become more involved in design)

  Capital-intensive facilities utilize incentives such as tax breaks, price breaks, and so on, which can influence the *make versus buy* decision

- **Political and economic factors**

  Trade protection mechanisms such as tariffs, quotas, voluntary export restrictions, local content requirements, environmental regulations, and government procurement policies (discount for local)

  Customs duties, which differ by commodity and the level of assembly

  Exchange rate fluctuations and operating flexibility

Global Supply Chain Strategy Development

Today, in most industries, it is necessary to develop a global view of your organization's operations to survive and thrive. However, many companies find it difficult to transition from domestic to international operations, despite the fact that there have been significant improvements in transportation and technology over the past 25 years.

To be successful in the global economy, a company must have a supply chain strategy. This should include significant investments in enterprise resource planning (ERP) and other supply chain technology to prepare them to optimize global operations by linking systems across their businesses globally, thus helping them to better manage their global supply chains.

Earlier in this book, we discussed organizational strategies and how the supply chain must support them. It is no different when discussing a global supply chain. In general, an organization should have their global supply chain set up to maximize customer service at the lowest possible cost.

Kauffman and Crimi, in their paper "A Best-Practice Approach for Development of Global Supply Chains" (2005), suggest that developing a global supply chain not only requires the same information as when developing one domestically, but also requires additional information on international logistics, law, customs, culture, ethics, language, politics, government, and currency. Cross-functional teams should be utilized that are supplied with detailed information, including the *what*, *when*, and *where* of the global supply chain as well as quantity demand forecasts. Supplier evaluations must include the ability for them to handle international operations and subsequent requirements.

To actually implement a global supply chain for your business, after identifying your supply chain partners, the team should document and test the required processes and procedures before implementing. All participants must be trained in the processes and procedures with metrics established to manage and control the global operations. The team must establish a project plan with responsibilities and milestones for the implementation.

The actual step-by-step approach for developing global supply chains recommended by Kauffman and Carmi is as follows:

1. Form a cross-functional global supply chain development team.

   Include all affected parties, internal and external.

   The team composition may change as development and implementation proceeds.

2. Identify needs and opportunities for supply chain globalization.

   Determine the requirements your supply chain must meet:

   Commodities, materials, services required... dollar value of materials and services... importance of commodities, materials, and services...

   Performance metrics for qualification and evaluation of suppliers.

   Determine the current status of your supply chain as is:

   Existing suppliers of materials and services

   Customers...

   Commodity markets...

   Current performance, problem areas

   Competitiveness...

   Fit of your current supply chain with your operational requirements.

   The main components of this particular framework... should include all operational dimensions of supply chains, which must be identified, considered, and included in any determination of requirements and assessment of current status of supply chains.

3. Determine commodity/service priorities for globalization consideration based on needs and opportunities.

4. Identify potential markets and suppliers and compare to markets, suppliers, and supply chain arrangements, operations, and results.

5. Evaluate/qualify markets and suppliers, identify supplier pool (determine best ones based on likely total cost of ownership (TCO), and best potential to meet or exceed expectations and requirements).

6. Determine selection process for suppliers (request for proposal [RFP], negotiation, and so on).

7. Select suppliers or confirm current suppliers.

8. Formalize agreements with suppliers.

9. Implement agreements.

10. Monitor, evaluate, review, and revise as needed.

Whatever your company's global strategy, it must be supported by a strong transportation network.

International Transportation Methods

The primary methods of international transportation are ocean and air between countries and motor and rail within overseas countries.

Ocean

Ocean transport is perhaps the most common and important global shipment method and accounts for approximately two-thirds of all international movements. Some of the advantages of this mode of international transportation are low rates and the ability to transport a wide variety of products and shipment size.

It breaks up into three major categories of 1) liner services, which have regular routes and operate to a schedule and operate as a common carrier, and 2) charter vessels, which are for hire to carry bulk (dry or liquid) or break bulk (cargoes with individually handled pieces) to any suitable port in the world, and 3) private carriers.

Air

International air transportation is primarily used for premium or expedited shipments due to its fast transit times. However, as explained previously, this mode is subject to high transportation rates.

Motor

When in a foreign country, like domestically, motor carrier is one of the most popular forms of transportation because its standardization reduces complexity. For example, motor transport is the primary form of transportation when shipping goods to between the United States and Mexico or Canada and is common in Europe. It also plays a major role in intermodal shipments, especially at ports when unloading container ships.

Rail

International railroad use is also highly similar to domestic rail use, and intermodal container shipments by rail are increasing.

Global Intermediaries

In addition to the global intermediaries such as freight forwarders and customs brokers discussed in Chapter 7, "Transportation Systems," there may be the need for storage and packaging expertise.

Storage Facilities

What are known as *transit sheds* can provide temporary storage while the goods await the next portion of the journey in a foreign land. In some cases, the carrier may provide storage on-dock, free of charge until the vessel's next departure date. Public warehouses are available for extended storage periods.

Bonded warehouses, mentioned in Chapter 8, "Warehouse Management and Operations," operate under customs agency's supervision and can be used to store, repack, sort, or clean imported merchandise entered for warehousing without paying import duties while the goods are in storage.

Packaging

Export shipments moving by ocean transportation typically require stricter packaging than domestic shipments because the freight handling involves many firms and the firms are located in different countries. As a result, the shipper may find settling liability claims for damage to export goods difficult.

Global Supply Chain Risks and Challenges

The global supply chain is fraught with risks and challenges.

As operations become more complex, logistics becomes more challenging, lead times lengthen, costs increase, and customer service can suffer. With a global footprint, different products are directed to more diverse customers via different distribution channels, requiring different supply chains.

There are many other, additional issues to address, such as the identification of sources capable of producing the materials in the quality and quantity required; the protection of a firm's intellectual property; import/export compliance issues; communication with suppliers and transportation companies; differences in time zones; language and technology; and product security while in transit.

Questions to Consider When Going Global

All of this raises some initial questions that companies need to consider as their operations globalize, as was pointed out in a PWC-MIT forum on supply chain innovation (PWC.com, 2013).

The questions and findings from the forum were as follows:

- What are the drivers of supply chain complexity for a company with global operations?

  Supply chains are exposed to both domestic and international risks. The more complex the supply chain, the less predictable the likelihood and the impact of disruption. Over recent years, the size of the supply chain network has increased, dependencies between entities and between functions have shifted, the speed of change has accelerated, and the level of transparency has decreased.

  Overall, developing a product and getting it to the market requires more complex supply chains, needing a higher degree of coordination.

- What are the sources of supply chain risk?

  Risks to global supply chains vary from controllable to uncontrollable ones and include the following:

  Raw material price fluctuation

  Currency fluctuations

  Market changes

  Energy/fuel prices volatility

  Environmental catastrophes

  Raw material scarcity

  Rising labor costs

  Geopolitical instability

- What parameters are supply chain operations most sensitive to?

  Respondents replied that their supply chain operations were most sensitive to reliance on skill set and expertise (31%), price of commodities (29%), and energy and oil (28%). For example, when U.S. diesel prices rose significantly in 2012, shippers rapidly adjusted budgets to offset the increased costs higher fuel prices produce.

- How do companies mitigate against disruptions?

  A great majority of respondents (82%) said they had business continuity plans ready. Nissan, for example, had a well-thought-out and exercised business continuity plan ready to kick into action to facilitate a quick recovery. Other major strategies by respondents included the following:

    Implement dual sourcing strategy

    Use both regional and global strategy

    Pursue (first- and second-tier) supplier collaboration

    Pursue demand collaboration with customers

Key Global Supply Chain Challenges

According to a survey by PRTM consultants for Supply Chain Digest (SCDigest Editorial Staff, 2010), key global supply chain challenges include the following:

- **Supply chain volatility and uncertainty have permanently increased:** Market transparency and greater price sensitivity have led to lower customer loyalty. Product commoditization reduces true differentiation in both the consumer and business-to-business (B2B) environments...

- **Securing growth requires truly global customer and supplier networks:** Future market growth depends on international customers and customized products. Increased supply chain globalization and complexity need to be managed effectively...

- **Market dynamics demand regional, cost-optimized supply chain configurations:** Customer requirements and competitors necessitate regionally tailored supply chains and product offerings. End-to-end supply chain cost optimization will be critical...

- **Risk management involves the end-to-end supply chain:** Risk and opportunity management should span the entire supply chain—from demand planning to expansion of manufacturing capacity—and should include the supply chains of key partners...

- **Existing supply chain organizations are not truly integrated and empowered:** The supply chain organization needs to be treated as a single integrated organization. To be effective, significant improvements require support across all supply chain functions.

Risk Management

An organization's supply chain is greatly impacted by globalization and its inherent logistical complexity. This has resulted in having risk beyond just the demand and supply variability, limited capacity, and quality issues that domestic companies have traditionally faced, to now include other trends such as greater customer expectations, global competition, longer and more complex supply chains, increased product variety with shorter lifecycles, and security, political, and currency risks.

As a result, it is important for global supply chain managers to be aware of the relevant risk factors and build in suitable mitigation strategies.

Potential Risk Identification and Impact

Before planning for risks in your supply chain, you must first identify potential risks and their impact.

To accomplish this, many companies use a *vulnerability map* or risk matrix to visualize unforeseen and unwanted events, as shown in Figure 11.1 (Sheffield & Rice, Jr., 2005).

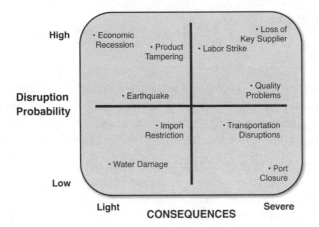

Figure 11.1 Vulnerability map

This type of analysis has two dimensions: disruption probability and consequences. Obviously, risks with a high disruption probability and severe consequences should be given a great deal of attention.

One problem with this method is that it relies heavily on risk perception, which can vary depending on recent events, a person's experience and knowledge, their appetite for risk, and their position in the organization, among other things.

Sources of Risk

Before determining a risk management strategy for your organization, it is important to consider the possible sources of risk. There are five sources of risk in a supply chain, some of which are internal, others external to your organization (see Figure 11.2) (Christopher & Peck, 2005).

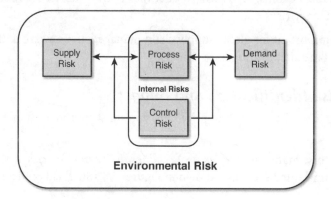

Figure 11.2 Sources of risk in the supply chain

Internal Risks

Process risk refers to the value-adding and managerial activities undertaken by the firm and to disruptions to these processes. These processes are usually dependent on internally owned or managed assets and on the existing infrastructure, so the reliability of supporting transportation, communication, and infrastructure should be carefully considered.

Control risks are the rules, systems, and procedures that determine how organizations exert control over the processes and are therefore the risks arising from the use (or misuse) of these rules. For the supply chain, they include order quantities, batch sizes, safety stock policies, and so on and any policies and procedures that cover asset and transportation management.

External Risks

Demand and supply risk are external to the organization, but are internal to the networks through which materials, products, and information flow between companies. The organization should consider potential disruptions to the flow of product and information from within and between all parties in the extended supply chain network and at least understand and monitor the potential risks that may affect other supply chain partners.

Supply risk is the upstream equivalent of demand risk and relates to potential or actual disturbances to the flow of product or information from within the network, upstream of your organization.

Environmental risks are disruptions that are external to the network of organizations through which the products flow. This type of event can impact your organization directly, on those upstream or downstream, or on the marketplace itself.

Environmental-related events may affect a particular product (for example, contamination) or place through which the supply chain passes (for example, an accident, direct action, extreme weather, or natural disasters). They may also be the result of sociopolitical, economic, or technological events far removed from your firm's own supply chains, with the effects often reaching other industry networks. In some cases, the type or timing of these events may be predictable (for example, regulatory changes), and many will not be, but their potential impact can still be evaluated (Christopher & Peck, 2005).

Supply Chain Disruptions

Supply chain disruptions are the actual occurrence of risks, including the categories mentioned earlier, and are unplanned and unanticipated events that disrupt the normal flow of goods and materials within a supply chain.

A triggering event usually happens, followed by the situation (with its consequences) that occurs afterward.

Disruptions that a company has to deal with come primarily, although not always as previously mentioned, from customers, suppliers, or the supply chain. The consequences can be immense to your company and can include higher costs, poor performance, lost sales, lower profits, bankruptcy, and damage to your organization.

The actual characteristics of the supply chain structure you have may determine the drivers of your supply chain's vulnerability.

These characteristics may including the following:

- Complexity of the supply chain (for example, global versus domestic sourcing).
- Density of the supply chain. (That is, using these high-density regions leads to higher vulnerability of supply chains.)
- Single or sole sourcing versus multiple vendors for the same item.
- Lean and just-in-time (JIT) production philosophies require precise timing.
- Centralization of warehouse/manufacturing locations results in lengthy lead times due to distance issues.
- Dependency on major suppliers/customers (that is, the all "your eggs in one basket" syndrome).
- Dependency on IT infrastructure, electricity, and so on.
- Flexible, secure supply chains with a diversified supplier base are less vulnerable to disruptions than those that are not.

Therefore, to a great degree, potential disruptions are the result of *conscious* decisions regarding how you design the supply chain. Risk management is about using innovative planning to reduce potential disruptions by preparing responses for negative events.

Risk Mitigation

Depending on the type of supply chain risk, what follows are some common supply chain risks and tactics for risk mitigation (Heizer & Render, 2013):

- **Supplier failure to deliver:** Use multiple suppliers with contracts with containing penalties. When possible, keep subcontractors on retainer.

  Example: McDonald's planned its supply chain many years before opening stores in Russia. All plants are monitored closely to ensure strong links.

- **Supplier quality failure:** Ensure that you have adequate supplier selection, training, certification, and monitoring processes.

  Example: Darden Restaurants (that is, Olive Garden restaurants) uses third-party audits and other controls on supplier processes and logistics for reduction of risk.

- **Logistics delays or damage:** Have multiple or backup transportation modes and warehouses. Make sure that you have secure packaging and execute contracts with penalties for nonconformance.

  Example: Walmart always plans for alternative origins and delivery routes bypassing problem areas when delivering from its distribution centers to its stores with its private fleet.

- **Distribution:** Have a detailed selection and management process when using public warehouses. Make sure that your contracts have penalties for nonconformance.

  Example: Toyota trains its dealers on improving customer service, logistics, and repair facilities.

- **Information loss or distortion:** Always backup databases within secure information systems. Use established industry standards and train of supply chain partners on the understanding and use of information.

  Example: Boeing utilizes a state-of-the-art international communication system that transmits engineering, scheduling, and logistics data to Boeing facilities and suppliers worldwide.

- **Political:** Companies can purchase political risk insurance. This is also the situation where you may decide to go the route of franchising and licensing with your business.

  Example: Hard Rock Cafe restaurants try to reduce political risk by franchising and licensing in countries where they deem that the political and cultural barriers are great.

- **Economic:** Hedging, the act of entering into a financial contract to protect against unexpected, expected, or anticipated changes in currency exchange rates, can be used to address exchange rate risk.

  Example: Honda and Nissan have moved some of its manufacturing processes out of Japan since the exchange rate for the yen has made Japanese-made automobiles more expensive.

- **Natural catastrophes:** In many cases, natural disasters can be planned for by taking out various forms of insurance (for example, flood insurance). Companies may also consider alternate sourcing for example.

  Example: Toyota, after the 2011 earthquake and tsunami, has established at least two suppliers, in different geographic regions, for each component.

- **Theft, vandalism, and terrorism:** Again, in some cases, there is insurance available for these types of risk. Companies also enforce patent protection and use security measures such as radio frequency identification (RFID) and Global Positioning System (GPS).

  Example: Domestic Port Radiation Initiative. The U.S. government has established radiation monitors at all major U.S. ports that scan imported containers for radiation.

One reason that risk exists in a supply chain, global or domestic, is that, due to its complexity, many companies choose to outsource many services. Risk of this type can be minimized if managed properly, which is the topic of the next chapter.

PART IV

Supply Chain Integration and Collaboration

Supply Chain Partners

T he supply chain and logistics function is always a prime candidate for outsourcing. Strategically speaking, most successful companies stay with their core competencies and let outside entities help with the rest.

This can range from sourcing of functional areas such as materials, transportation/warehouse services, and manufacturing, to most of an organization, known as a *virtual company*.

There are actually four major ways to get things done in business:

- **Internally:** Processes that are core competencies are usually the best way to perform an activity.

- **Acquisition:** Gives the acquiring firm full control over the way the particular business function is performed. Can be difficult and expensive (culture/competitors).

- **Arm's-length transactions:** Most business transactions are of this type. These are short-term arrangements that meet a particular business need but don't lead to long-term strategic advantages.

- **Strategic alliances:** Longer-term multifaceted partnerships between two companies that are goal oriented. There are both risks and rewards to an alliance, which are shared, but alliances can lead to long-term strategic benefits for both partners. Strategic alliances in the supply chain include third-party (3PL) and fourth-party (4PL) logistics services.

Outsourcing

Outsourcing is the contracting out of a business process to a third-party where an organization transfers some internal activities and resources of a firm to outside vendors. It is really an extension of the subcontracting and contract manufacturing of product, which have both existed for a very long time. Outsourcing includes both foreign and domestic contracting, and can include *offshoring* (that is, relocating a business function to another country).

Most firms outsource some functions where they don't feel that they have a competency, such as the fulfillment of orders, as in the printing industry, or using for-hire motor carriers for delivery to customers, which is common in many industries.

In the 1990s, to reduce costs, companies began to outsource a variety of services, such as accounting, human resources, technology, internal mail distribution, security, and facility maintenance.

From a supply chain standpoint, a variety of functions may be candidates for outsourcing, such as warehousing, transportation, freight audit and payment, procurement, and customer service/call centers.

In today's global economy, organizations look for long-term strategic partnerships for functions and services, some that might even be considered *core competencies,* to gain a strategic advantage. The rapid increase in outsourcing can at least partially be attributed to increased technological expertise, more-reliable and less-costly transportation service, and advancements in telecommunications and computer systems.

Reasons to Outsource

Reasons to outsource include the following:

- **Lower operational and labor costs:** These are usually the primary reasons why companies choose to outsource. When properly executed, it has a defining impact on a company's revenue and can deliver large savings.

- **Company focus:** So that a company can continue to focus on core business processes while delegating less-important, time-consuming processes to external partners.

- **Knowledge:** It can enable companies to leverage a global knowledge base and have access to world-class capabilities.

- **Freeing up internal resources:** They can be put to more effective use for other purposes.

- **Access to resources not available internally:** Companies may have internal resource constraints.

- **Specialists for hard-to-manage areas:** By delegating responsibilities to external agencies, companies can hand off functions that are difficult to manage and control while still realizing their benefits.

- **Risk mitigation:** Outsourcing, and especially offshoring, helps companies to mitigate risk.

- **Reengineering:** Can enable companies to realize the benefits of a reengineering process.

- **New markets:** Some companies may outsource to help them expand and gain access to new market areas, by moving the point of production or service delivery closer to their end users.

Steps in the Outsourcing Process

Outsourcing, which in many ways is similar to any sourcing (of goods or services) initiative, must follow a distinct process to be successful. A good example of one methodology is encompassed in the following steps:

1. **Plan initiatives.**

   Establish cross-functional teams to assess the risks and resources. The team sets objectives, deliverables, and timetables and is responsible for achieving critical management buy-in.

   This step should also include sharing the information with employees. If not, employees may assume the worst, causing a lowering of morale.

2. **Explore strategic implications.**

   This is the step in which outsourcing is used as a strategic tool where one examines current and future organizational structures and considering current and future core competencies of an organization. It is a long-term view to see whether the solution is a good fit.

3. **Analyze cost and performance.**

   The organization must next make sure that all costs needed to support the activity, direct and indirect, are considered. Current performance must also be measured and analyzed to establish a baseline against which to measure improvement.

4. **Select providers.**

   Finding potential providers can be done in a variety of ways, including the use of references from business associates, directories, advertisements, RFIs (request for information), and so on.

   After you have narrowed the list of potential outsource partners, you'll need to develop and send out requests for proposals (RFPs). The RFPs should at least include what is required of the outsourcer in the way of both services and information. Once returned, RFPs should be evaluated in terms of both qualifications and cost.

5. **Negotiate terms.**

   You should next map out with the provider the services to be provided and pricing (including how changes in scope or volume will be handled), as well as performance standards, management issues, and transition and termination provisions. A clear, well-documented understanding during this process can contribute greatly to the success of the relationship. The organization also needs to consider worst-case scenarios, which outline a plan of action should the outsourcing relationship fail.

6. **Transition resources.**

One of the biggest challenges to the staff of the organization is managing the impact of the potential change and the actual transition. Open communication from the start is critical.

Human resource issues should be carefully addressed, and with sensitivity. Any staff that are about to be terminated because their jobs have been outsourced should be treated with sensitivity and respect. This will have an impact on how the remaining employees, who were not outsourced, contribute in the future. So, providing terminated employees such benefits as outplacement services is a good idea when possible.

7. **Manage the relationship.**

Outsourcing requires a different set of skills, as besides scheduled meetings and reports, unforeseen things may come up. So in addition to monitoring performance and evaluating results, a relationship of trust that enables problem solving is critical with your outsourcing partner (Greaver, 1999).

Supply Chain and Logistics Outsourcing Partners

As supply chain and logistics is an area commonly considered for outsourcing, there are a variety of options available to an organization. We will examine some of the major options below.

Traditional Service Providers

In the supply chain and logistics function, the two traditional service providers are in the area of transportation and warehousing, the characteristics of which were discussed earlier in this book.

To recap, the for-hire transportation industry has thousands of carriers who specialize in product movement between geographic locations who provide an assortment of services in various modes of transportation with related technology.

For-hire transportation companies offer specialization, efficiency, and scale economies to their customers. The choice for users of these services is to invest in and operate the vehicles themselves or use for-hire services at a negotiated (or standard) rate.

Public warehouses also offer storage and value-added services on a contracted basis, with the customer having similar choices as was mentioned for transportation (that is, invest capital or "pay as you go").

The main benefits of using public versus private warehousing is that there is no capital investment required for the building and equipment (and no employees to hire and manage) and the potential to consolidate small shipments with products of other firms that use the same public warehouse for combined delivery at a lower transportation rate. The warehouse charges are both time (that is, storage) and/or transaction (that is, handling and specialized services) based, as stated in the contract.

Third-Party Logistics Providers

A third-party-logistics provider (3PL) is an external supplier that performs all or part of its customers outsourced logistics functions and is usually asset based (although not always, as in the case of a financial, forwarder, or information systems-based firm).

Most typically, 3PLs specialize in integrated operation, warehousing, and transportation services based on its customers' needs, which may vary based on market conditions and delivery service requirements for their products and materials.

In the 1970s and 80s, mostly operational, repetitive, transactional operations were outsourced in the logistics function that required use of the service provider's transportation management system (TMS) and warehouse management system (WMS) (see Figure 12.1).

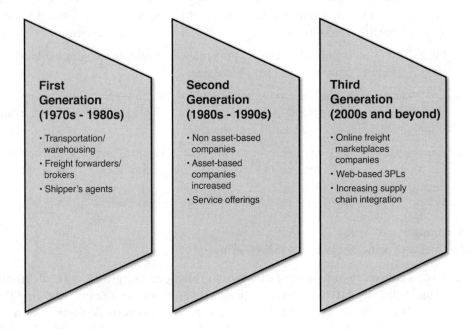

Figure 12.1 Evolution of third-party logistics service providers

By 2009, the 3PL industry had grown to almost $107 billion in size, and in addition to basic warehouse and transportation services, many 3PLs now offer value-added services related to the production or procurement of goods such as order fulfillment, labeling, packaging, assembly, kitting, reverse logistics, information technology services, customs brokering, cross-docking, and forwarding.

As a result of its far-reaching impact, the relationship between an organization and a 3PL vendor is a strategic, long-term, multifunctional partnership.

Advantages and Disadvantages of a 3PL

The use of 3PLs can offer both advantages and disadvantages. It is up to each organization to consider the risks and rewards before proceeding.

Advantages

There are many advantages to using 3PL service providers, including the following:

- **Focus on core strengths:** Allows a company to focus on its core competencies and leave logistics to the experts.
- **Provides technological flexibility:** Technology advances are adopted by better 3PL providers in a quicker, more cost-effective way than doing it yourself. 3PLs may already have the capability to meet the needs of a firm's potential customers.
- **Flexibility:** The use of a 3PL offers companies flexibility in geographic locations, service offerings, resources, and workforce size.
- **Cost savings:** 3PLs offer the economic principle of specialization by building up logistical infrastructures, methodologies, and computer-based algorithms to maximize shipping efficiency to cut a client's logistics costs.
- **Capabilities:** Smaller companies have to make large investments to expand their logistic capabilities. It may be more cost-effective and quicker to add capabilities through 3PLs.

Disadvantages

Disadvantages of using 3PLs include the following:

- **Loss of control in outsourcing a particular function:** As most 3PLs are on the outbound side, they heavily interact with an organization's customers. Knowing that, many 3PL firms work very hard to address these concerns by doing things such as painting client company logos on the sides of trucks, dressing 3PL employees in the uniforms of the hiring company, and providing extensive reporting on each customer interaction.

- **Pricing models:** By handing logistics over to a 3PL service, a company may be missing the possibility that an in-house logistics department could come up with a cheaper and more efficient solution.

- **Dependency:** If a 3PL is not working out as expected, switching a company's logistical support can cost the company a great deal in unanticipated costs resulting from the changes in pricing or unsatisfactory service reliability from the 3PL service.

- **Logistics is one of the core competencies of a firm:** In this case, it makes no sense to outsource these activities to a supplier who may not be as capable as the firm's in-house expertise.

Example

Ryder is one of the largest and most recognizable 3PL brand names. They are a lead logistics provider for most General Motors plants and services, Chrysler/Fiat, Toyota, and Honda, plus a multitude of tier-one suppliers. Among their services, they run inbound supply chain management, sequencing centers, and just-in-time (JIT) and dedicated contract carriage operations for clients.

Results that Ryder has had with clients include the following:

Apria Healthcare: In 2012, Apria contracted with Ryder Supply Chain Services (SCS) to provide dedicated contract carriage (DCC) dry-van truckload transportation services for products moving from its seven distribution centers (DCs) and cross-dock to its branch operations. As part of the operation, Ryder SCS also manages unattended deliveries, hazardous materials, product segregation, and vendor returns.

The following actions were taken:

The majority of inbound supplier shipments consolidated onto full truckloads. (More than 75% shipments now move at truckload rates.)

Supplier shipment frequency reduced to one to two times per week to each Apria DC.

Expedited freight greatly reduced, and the standard shipment method is now truckload and less than truckload.

The network has been optimized by filling Ryder's dedicated operation's backhaul lanes with inbound shipments from suppliers to Apria DCs.

Carrier Corporation (Mexico): Ryder supports three Carrier air conditioner-related operations in and around Monterrey.

At the Carrier residential air conditioner factory in Monterrey, Ryder has 110 employees integrated with 1,100 Carrier employees. Ryder's personnel handle receiving, manufacturing (JIT/Kanban), support, and shipping. Ryder does all phases of the materials management for Carrier, including sequencing, kitting, picking, and packing. Ryder also handles a large portion of inbound material with dedicated and managed transportation for the facility (Armstrong & Associates, Inc., 2007 and 2013).

Fourth-Party Logistics Service Providers

A fourth-party logistics service provider (4PL) is an integrator that assembles not only the resources (including possibly 3PLs) but also the planning capabilities and technology of its own organization and other organizations to design, build, and run comprehensive supply chain solutions for clients.

This is as opposed to a 3PL service provider that typically targets a single function. A 4PL targets management of the entire process. In some ways, a 4PL can be thought of as a general contractor that manages other 3PLs, transportation companies, forwarders, custom house agents, and so on, and therefore takes responsibility of a complete process for the customer (see Figure 12.2).

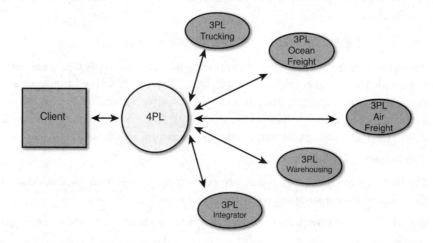

Figure 12.2 Fourth-party logistics service provider

The key difference between 4PL and other approaches to supply chain outsourcing is a unique ability to deliver value to client organizations across the entire supply chain.

4PL service providers are able to combine multiple clients spend to take advantage of high-volume discounts, which in turn enables them to provide affordable services to businesses of all sizes, ranging from small e-commerce start-ups to multinational manufacturers.

4PLs come in many varieties. For example, UPS now offers 4PL service that includes global supply chain design and planning, logistics and distribution, customs brokerage and international trade services, as well as freight services via ocean, air, or ground.

This is opposed to a 3PL service provider, which usually focuses on one or two areas of expertise, such as warehousing, distribution, or freight forwarding, often resulting in multiple 3PLs being used to complete the supply chain.

Players

The players involved in creating a 4PL organization are as follows:

- The client, who provides start-up equity, some assets, working capital, operational expertise, staff, and, of course, procures logistics services from the 4PL organization
- 3PL service providers (primarily for transportation services and distribution facilities)
- The 4PL partner, who may provide a range of resources, including logistics strategy, reengineering skill, benchmarks, IT development, customer service, and supplier management and logistics consulting

The typical 4PL organization is hybrid, in that it is formed from a number of different entities and typically established as a joint venture (JV) or long-term contract.

The goals of partners and clients are aligned through profit sharing, and the 4PL is responsible for the management and operation of the entire supply chain, with a continual flow of information between partners and the 4PL organization.

Components Required for Success

The components required for a successful 4PL strategy are as follows:

- **Leadership:** Must be a bit of a supply chain visionary and deal maker with a multiple-customer relationship. This component acts as the project manager as well as a service, systems, and information integrator.
- **Management:** Experienced in logistics operations, optimization, and continuous improvement to run day-to-day operations and make important decisions. This component must also manage multiple 3PLs.
- **Information technology (IT) support:** The "brains" of the operation, with full integration and support of all systems in the supply chain.
- **Assets:** Transportation and warehouse assets as well as outsourced contract manufacturing and co-packing and procurement services.

Example

Menlo Worldwide Logistics (www.con-way.com) is a leader in 4PL that specializes in the integration of all functions across the supply chain, from sourcing of raw materials, through product manufacturing, to the distribution of finished goods.

Menlo acts as a neutral single point of control for your supply chain by managing the procurement, optimization, information analytics, and operations of your supply chain network. They help their clients to create flexible supply chain solutions that support their corporate strategy while increasing supply chain savings and service improvements. They act as a change agent to ensure the success of a client's supply chain transformation. They provide the following:

- Deployment of Lean tools and methodologies
- Self-funding initiatives
- Delivery of flexible supply chains built to withstand business change and improve velocity
- Delivery of best-of-breed and customer-specific business solutions

To get an idea of what kind of success companies can have implementing a 4PL strategy, here are some results from some Menlo clients (Con-way, 2014):

- **Automotive customer:** Managed more than 12,000 locations, $4 billion logistics spend with $648 million in savings. Utilized business case methodology to identify and measure savings.
- **High-tech customer:** More than $30 million in savings throughout engagement, $9 million cost reduction through network rationalization and optimization in year one, and integration of regional operations into enterprise-wide network.
- **Heavy equipment customer:** Support $400 million global logistics network, on track to achieve a 25% reduction in supply chain spend. Cross-business-unit solutions, including the following:

  Global transportation networks

  Regional infrastructure requirements

  New landed cost modeling

Many of the same career opportunities covered in the transportation and warehousing chapters are also available with 3PLs and 4PLs. These include operations, management, consulting, and sales. The field has grown significantly in the past 25 years, and so it is a great source for supply chain and logistics careers.

Next, we will discuss collaborative relationships that are primarily between the major players in the supply chain—retailers, distributors, manufacturers, and suppliers—who are primarily focused on the sharing of information to improve planning and management of inventory throughout the entire supply chain.

13

Supply Chain Integration Through Collaborative Systems

Supply chain integration refers to the degree to which the firm can strategically collaborate with their supply chain partners and collaboratively manage the intra- and interorganization processes to achieve the effective and efficient flows of product and services, information, money, and decisions (see Figure 13.1). The objective of this integration is to provide the maximum value to the customer at low cost and high speed.

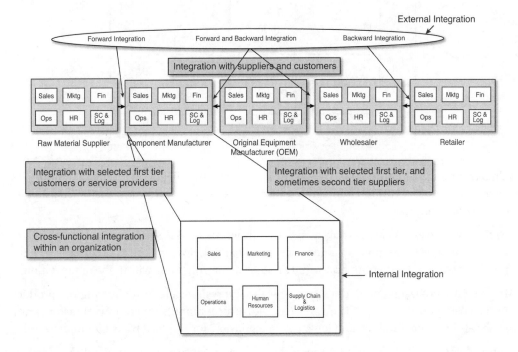

Figure 13.1 Major types of integration

Integration is not the same as collaboration, as integration is the alignment and linking of business processes and includes various communication channels and connections within a supply network. Collaboration, in contrast, is a relationship between supply chain partners that is developed over time. Integration is possible without collaboration, but it can be an enabler of collaboration.

There are two general categories of business integration: internal and external, which we will now explore.

Internal and External Integration

Internal Integration

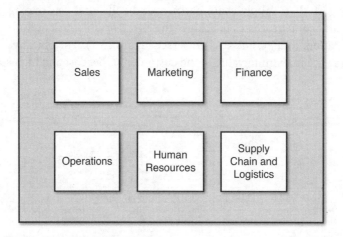

Understanding the entire supply chain of an organization begins with understanding its internal processes, because an integrated firm presents a united front to customers, suppliers, and competitors.

As the saying goes, "A chain is only as strong as its weakest link." So, it is critical that there are good communications, policies, and procedures that link not only the internal supply chain processes with each other but also with the other major functions within the organization.

In general, purchasing, operations, and logistics are responsible for delivering the product to the customer. Purchasing is a gatekeeper for process inputs, operations transform raw materials into final product, and logistics is responsible for physical transfer and delivery.

Internal supply chain integration therefore is a process of interaction and collaboration in which manufacturing, purchasing, and logistics work together in a cooperative manner to arrive at mutually acceptable outcomes for their organization.

It is important to integrate communications and information systems so as to optimize their effectiveness and efficiency, and this can be achieved by structuring the organization and the design and implementation of information systems where non-value-adding activity is minimized; costs, lead times, and functional silos are reduced; and service quality is improved.

Many organization's today use process improvement tools such as Lean and Six Sigma (and the combination of the two known as *Lean Six Sigma*, covered later in this book) to analyze existing organizational structures, eliminate non-value-adding activities, and implement new work structures so that the organization is optimally aligned.

An integrated enterprise resource planning (ERP) is a key enabler of internal integration, often exposing remaining non-value-added activities in the organization and allowing for better communication and collaboration through a common database.

Internal integration needs more than a system and proper organization. It also needs the following:

- **Shared goals:** Which refers to the extent to which the manager of each key function (purchasing, operations, and logistics) is familiar with the strategic goals of each of the other two focal functions
- **Cooperation:** Measured by the frequency of requests from other focal functions fulfilled by the members of each focal function
- **Collaboration:** Defined as the frequency at which a member of a key function actively works on issues with members from the other key functions

External Integration

External or interorganization integration involves the sharing of product and service information and knowledge between organizations in a supply chain. Like internal integration, it also requires shared goals, cooperation, and collaboration to work successfully.

This enables all stages of the supply chain to take actions that are aligned and increase total supply chain surplus (that is, the difference between revenue less cost to produce and deliver product to the customer).

It requires that each stage in the entire supply chain share information and take into account the effects of its actions on the other stages.

If the objectives of the different stages conflict with each other or information moving between stages is delayed or distorted, lack of coordination will result, resulting in the bull-whip effect described earlier in Chapter 2, "Understanding the Supply Chain."

Successful collaboration relies on the development of mutual trust between you and your partners, as well as the willingness to share information (electronically and manually) that can benefit all the members of your collaborative team. The goal is to treat all suppliers, outsourcing partners, customers, and service providers as an extension of your organization.

Supply Chain Collaboration by Industry

Many industries are experimenting with supply chain collaboration, adapting the concept to fit their specific needs.

For example, consumer products and retail companies are implementing safety stock levels across their entire supply chains. Using point of sale (POS) and other information sources, these companies have increased service levels all the way to the store level.

Pharmaceutical and automotive industries have used collaboration to prevent counterfeit products from getting into their supply chains. For example, pharmacies are using radio frequency identification (RFID) to better manage the shelf life of perishable coded products.

High-tech space are looking to get production visibility beyond purchase-level response, to better control quality, cost, and availability to improve measurements of customer service such as customer request date.

Capital equipment and manufacturing companies are leveraging collaboration technologies to extend Lean supply chain principles across the enterprise by extending electronic kanban processes (that is, Lean technique to visually replenish inventory based on downstream demand) to suppliers.

Levels of External Collaboration

Across all industries, supply chain collaboration operates at the strategic, tactical, and execution level:

- **Strategic:** At this level, organizations and their partners make joint decisions on strategic issues such as production capacities, product design, production facility and fulfillment network expansion, portfolio joint marketing, and pricing plans.
- **Tactical:** This level involves sharing information with partners on topics such as forecasts, production, and transportation plans and capacities, bills of material, orders, product descriptions, prices and promotions, inventory, allocations, product and material availability, service levels, and contract terms such as supply capacity, inventory, and services.

- **Execution:** At this level, organizations and their partners engage in an integrated exchange of key transactional data such as purchase orders, production/work orders, sales orders, POS information, invoices, credit notes, debit notes, and payments (SAP, 2007).

Types of External Collaboration

External collaboration can range from relatively simple to very complex based on the amount of dependency and information sharing between parties (see Figure 13.2).

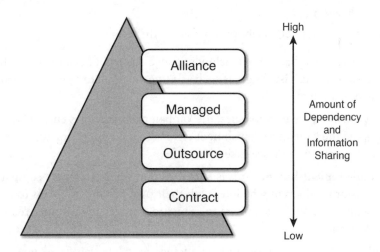

Figure 13.2 Range of external supply chain relationships

The simplest form of collaboration is contracting, which adds a time dimension to traditional buying and selling (that is, having price, service, and performance expectations over specific period).

The next level of collaboration is outsourcing, which shifts the focus from just buying materials to actually performing a specific service or activity.

After that comes what could be referred to as *managed* types of collaboration, where a dominate company uses a command and control system to direct the partner and there is limited sharing of strategic information and limited joint planning. This type of relationship has no specific time frame to it (that is, termination/rebid).

The most advanced forms of collaboration are alliances such as we've seen with Dell and its suppliers in a just-in-time (JIT) environment, where the parties voluntarily work together both strategically and operationally and there is an integration of human, financial, operational, and technical resources. There is an extensive amount of joint planning and anticipation of a long-term relationship.

As companies move along this path toward alliances, trust is a key component. Trust does not occur overnight; it requires ongoing interaction among organizations. It is necessary to first see reliability in operations and then a gradual sharing of all information for the relationship to function properly. Trust can be maintained by being open and honest with regard to key decisions (often referred to somewhat tongue in cheek as *opening your kimono*).

It is helpful to look at the types of external collaboration in terms of those with both suppliers and customers and the importance of integrating the sales & operations planning (S&OP) process to include information from both sources.

Supplier Collaboration

Some of the types of supplier collaboration include the following:

- **Kanban:** A signal-based replenishment process used in Lean or JIT production that uses cards or other visual signals such as a line on a wall to signal the need for replenishment of an item.

  Using collaborative technologies, the kanban process allows customers to electronically issue the kanban replenishment signals to their suppliers, who can then determine requirements and see exceptions.

- **Dynamic replenishment:** This is a process that where suppliers compare customer forecasts or production schedules with their own production plans to match supply and demand. It allows suppliers to adjust to changes in customer requirements or supply shortages.

- **Invoicing processes:** Automating invoicing and related processes gives the visibility to the vendor for the entire supply side, including purchase orders, releases, supplier-managed inventory, kanbans, and dynamic replenishment.

- **Outsourced manufacturer collaboration:** When managing outsourced manufacturing relationships or contract manufacturers, you must shift your focus from owning and organizing assets to working collaboratively with partners.

The collaborative efforts should help simplify processes such as product development and reduce manufacturing costs and improve reaction to response to customer demand.

Any efforts to automate these processes should support information sharing, collaboration, and monitoring activities that are needed to effectively manage the relationship with a contract manufacturer.

Customer Collaboration

Customer collaboration involves the receiving demand signals and automatically replenishing the customer's inventory based on actual demand. This is seen primarily in consumer products and other industries that have downstream distribution systems that extend to retailers.

This type of integration and collaborative effort enables manufacturers to shift from a *push* system to a demand *pull* supply chain while combining both forecasts and actual customer demand.

Collaborative replenishment processes are more responsive than purely forecast-based processes, and because they are driven largely by actual customer demand and also provide visibility in out-of-stock situations, manufacturers and retailers can react more quickly. Several of these are discussed later in the chapter and go by the names *quick response* (QR) and *efficient consumer response* (ECR). POS information can add visibility across the entire supply chain, as well, when included in a collaborative replenishment process.

Another type of customer collaboration that focuses on forecasts is known as collaborative planning, forecasting, and replenishment (or *CPFR*, which is a trademark of the Voluntary Inter-industry Commerce Standard Association [VICS]). It is an outgrowth from some of the earlier customer replenishment initiatives such as QR and ECR.

In general, CPFR is an attempt to reduce supply chain costs by promoting greater integration, visibility, and cooperation between trading partners' supply chains. It combines the intelligence of multiple trading partners in the planning and fulfillment of customer demand.

Figure 13.3 shows collaborative or vendor-managed inventory configurations in terms of the level of sophistication or complexity. Levels 1 and 2 have been implemented in various industries and would include programs such as QR and ECR. Levels 3 and 4 are more advanced and would include CPFR-like programs.

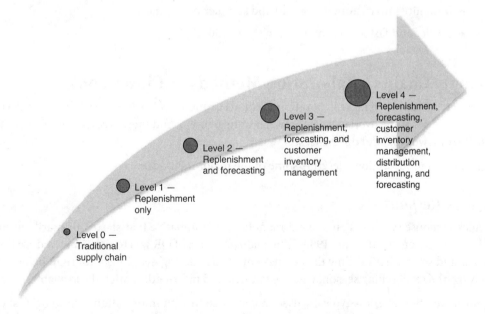

Figure 13.3 Types of collaborative or vendor-managed inventory in supply chains

Sales & Operations Planning

S&OP, discussed in Chapter 4, "Inventory Planning and Control," allows you to introduce collaborative information into the decision-making process, and when used as part of your collaborative efforts, it enables better communications between cross-functional groups and trading partners, both customers and suppliers.

Globalization and outsourcing generate a lot of information that can impact an organization's decision-making process externally to the enterprise. As a result of this, a comprehensive S&OP process is even more important. The S&OP process makes sure that your business is continually managed to meet organizational strategies, goals, and commitments, despite ongoing changes in your environment.

Benefits to Collaboration

Increased connectivity and collaboration between companies and their trading partners creates many benefits for both suppliers and your customers, such as the following:

- Higher inventory turns
- Lower fulfillment (transportation and warehousing) costs
- Lower out-of-stock levels and improved customer service
- Shorter lead times
- Early identification of changes to demand and improved market intelligence
- Visibility into customer demand and supplier performance
- Earlier and faster decision making (SAP, 2007)

Supply Chain Collaboration Methods: A Closer Look

A variety of supply chain collaboration methods or models have been used during the past 25 years or so. Some overlap exists between some of them, and there is some confusion as to their (somewhat subtle) differences.

The following subsections describe some of the major methods.

Quick Response

Quick response (QR) was an apparel manufacturing initiative that started primarily in the United States during the mid-1980s. The main objective of QR was to drastically reduce lead times and setup costs to allow the postponement of ordering decisions until right before, or during, the retail selling season, when better demand information might be available.

Implementation of QR is typically used in conjunction with information technologies such as electronic data interchange (EDI; a standardized format for businesses to exchange data electronically), barcodes, and RFID (the wireless use of radio frequency signals to transfer data to identify and track tags attached to objects).

A QR strategy can result in efficiencies such as maximized diversity of offering, quicker deliveries, faster inventory turns, fewer stock-outs, fewer markdowns, and lower inventory investment.

This type of strategy can also reduce the time between the sale and replacement of goods on the retailer's shelf because it places an emphasis on flexibility and product velocity to meet the changing requirements of a highly competitive and dynamic marketplace.

Efficient Consumer Response

Efficient Consumer Response (ECR), launched in 1984, is a grocery sector joint trade and industry organization with the goal of making the industry more responsive to consumer demand and to remove unnecessary costs from the supply chain.

The thinking is to improve the efficiency of a supply chain as a whole beyond the wall of retailers, wholesalers, and manufacturers, so that they can gain larger profits than each pursuing their own business goals.

One of the main practices used in ECR is to place smaller orders more often to shorten lead times, improve inventory turns, and reduce stock outs (see Figure 13.4). ECR is similar to QR except that it is targeted toward the grocery industry, where the supplier takes responsibility of monitoring and replenishing the retailer's distribution center inventories with approval of the retailer.

In the 1990s, when I was with Church and Dwight (Arm & Hammer products), we successfully implemented an ECR program to both place orders for and manage the inventory of our products at the distribution centers of a number of our grocery clients, including Wakefern (that is, ShopRite) and H. E. Butt.

Like QR, ECR is highly dependent on technology, using tools such as EDI, forecasting, and distribution requirements planning (DRP) software and point-of-sale (POS) data to manage the process.

The use of sophisticated technology like this as well as the lack of capabilities (both skill and technology related), the resistance of wholesalers, retailers, and manufacturers toward collaboration, and the attitudes of company personnel can all be barriers to a successful implementation of an ECR (or QR) type of program. It is well worth the effort, though, and can prove to be a *win-win* for all involved, because forecast accuracy tends to improve for the manufacturer and the retailer is relieved of managing the replenishment of some of its over 50,000+ stock keeping units (SKUs) while reducing stock-outs and inventory and ordering costs.

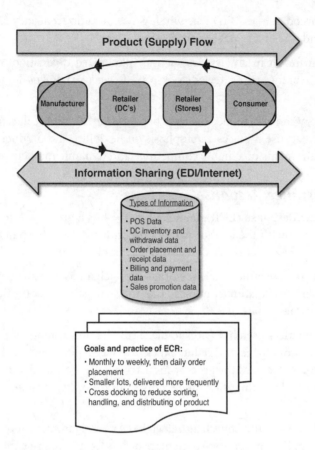

Figure 13.4 Efficient consumer response

Collaborative Planning, Forecasting, and Replenishment

Collaborative planning, forecasting, and replenishment (CPFR) is a form of collaboration that combines knowledge and information from multiple trading partners to reduce supply chain costs and improve efficiencies by linking sales and marketing best practices to supply chain planning and execution processes. Its overall objective is to increase customer service while reducing inventory, transportation, and logistics costs.

CPFR has its roots in efficient consumer response (ECR) and is an attempt to improve marketing, production, and replenishment functions, resulting in increased value to the consumer while at the same time improving supply chain performance for producers and retailers.

The Voluntary Inter-industry Commerce Solutions (VICS) association has defined both a framework and guidelines for CPFR, including elements for strategy and planning used for the development of joint business plans and the forms of collaboration, supply chain management focusing on forecasting and order planning, execution for the fulfillment of replenishment orders, and analysis for exceptions and performance metrics.

The benefits of CPFR can include the following:

- Improved forecast accuracy
- Smoother ordering patterns
- Increased sales revenues
- Higher order-fill rates
- Decrease in safety stock inventory levels
- Reduction in cost of goods sold (COGS) as a result of improved visibility of end-consumer demand, more accurate forecasts, and more stable production schedules

Similarly to QR and ECR, CPFR requires the use of technology, which can be shared technology, for planning, execution, and measurement.

As in most cases of intercompany integration and collaboration, besides having sophisticated supply chain processes and technology, the key is to have the proper organization alignment and training (see Figure 13.5).

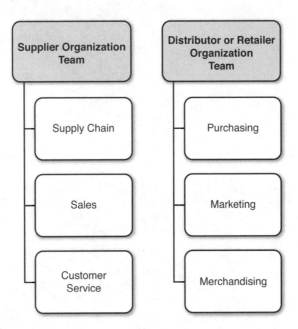

Figure 13.5 CPFR teams

On the supplier side, there will need to be CPFR teams made up of supply chain, sales, and customer service employees. On the customer side, which can be either the distributor or retailer, representatives from the purchasing, marketing, and merchandising functions will also need to be organized into a focused CPFR team.

Internal processes such as S&OP will have to be integrated with the CPFR process, and education initiatives are needed to make sure everyone understands and buys in to CPFR and its processes, as well as the implications of the coming change, benefits of CPFR, and the strategic importance of this type of initiative (Andrews, 2008).

Now that you understand how technology is a critical enabler of an effective supply chain (as discussed throughout the whole book), it is now time to take a closer look in the next chapter.

14

Supply Chain Technology

There really isn't any aspect of supply chain and logistics that isn't touched by technology in today's world. Thanks to both advances in software and hardware technologies and the Internet, companies of all sizes can automate and integrate internal processes and connect with customers and suppliers with ease.

Up to this point, we have briefly discussed technology applications in various aspects of supply chain and logistics management, including forecasting, inventory planning, production scheduling, and beyond. In this chapter, we go into a bit more depth in terms of understanding both the information flows and the systems used to make decisions at all levels of an organization, from longer-term decisions such as where to locate plants and warehouses, down to short-term decisions such as how many cases of product to ship to an individual warehouse on a given day.

Supply Chain Information

First, it should be made clear that there are subtle differences between data and information. Data are the facts from which information is derived to make decisions. Pieces of data are rarely useful alone; for data to become information, it needs to be put into context. That is the purpose of information systems.

According to Simatupang and Sridharan (2001):

> An interactive view of information enables people to define the level of information they need to solve problems or make decisions. Depending on the decisions, some people can use data to answer the questions, but others need to extract information from the same data to solve their problems. This interactive view also enables people to trace the source of knowledge from the available data, or to specify the required data based on their explicit knowledge (see Figure 14.1).
>
> An information system is used to collect, process, and disseminate information to make it available for decision makers at the right time. Traditionally, an information system deals with transferable data through plain media of communication such as

EDI and the Internet. The recent advance of information technology offers a rich variety of media such as video conferencing and online decision support systems that enable decision-makers to convert tacit knowledge into explicit knowledge and to share explicit knowledge.

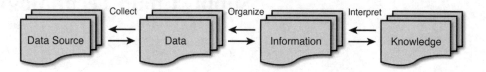

Figure 14.1 Interactive view of information

An organization's information requirements in general are that it needs to be easy to access, relevant to them, accurate, and timely.

Thus, the information technology used will have a direct impact on a company's performance, both internal and external through integration, which will enable collaboration (see Figure 14.2).

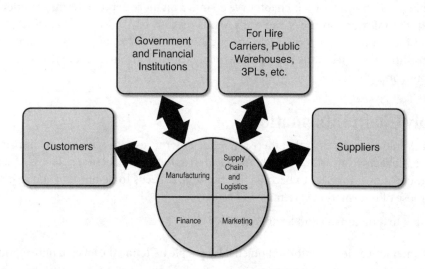

Figure 14.2 Internal and external supply chain information flows

The bullwhip effect, which was discussed earlier, is largely the result of poor information management in the supply chain and therefore can lead to excess inventory levels. By having greater demand visibility throughout the supply chain, inventory levels can be reduced. As mentioned earlier, it is possible to substitute information for excess inventory through the use of information systems.

McDonnell, Sweeney, and Kenny (2004) proposed four supply chain information technology solution categories:

- **Point solutions:** Used to support the execution of one link (or point) in the chain (for example, warehouse management systems [WMSs]).
- **Best of breed solutions:** Where two or more existing standalone solutions are integrated, usually using what is known as middleware technology to connect them.
- **Enterprise solutions:** Based on the logic of enterprise resource planning (ERP), these solutions attempt to integrate all departments and functions across a company into a single computer system that can serve all those different departments' particular needs.
- **Extended enterprise solutions (XES):** Refers to the collaborative sharing of information and processes between the partners along the supply chain using the technological underpinnings of ERP.

Since the 1990s, there has been movement away from point solutions toward enterprise solutions, partially reflecting the shift from internal and functional to an external management orientation.

This has been at least partially driven by other technologies, primarily electronic data interchange (EDI) and the Internet, which have enabled supply chain partners to use common data. This enables supply chain partners to act on actual demand, thus reducing the negative bullwhip effect to some degree.

Supply Chain Information Needs

SCM information systems use technology to more effectively manage supply chains.

Because there are so many applications in today's global supply chain, it is best to look at the technology from strategic, tactical, and execution viewpoints, as suggested by Bozarth and Handfield (2008). The viewpoints are as follows:

- **Strategic:** Develop long-term decisions that help to meet the organization's mission and focus on strategic plans for meeting such as new products or markets as well as facility capacity decisions.
- **Tactical:** Develop plans that coordinate the actions of key supply chain areas customers and suppliers across the tactical time horizon. They focus on tactical decisions, such as inventory or workforce levels. They plan, but don't carry out, the actual physical flows.
- **Routine:** Support rules-based decision making. Usually in short time frames, where accuracy and timeliness are important to the user.

- **Execution:** Typically more transactional oriented, where they record and retrieve transaction processing data and execute control physical and financial information flow. These systems usually have very short time frames, are highly automated, and use standardized business practices.

All these types of supply chain systems may also link both downstream with customers and upstream with suppliers.

Supply Chain Software Market

Today, most companies have implemented at least some components of supply chain systems, such as warehouse management or forecasting. The organizations that have taken an integrated, extended supply chain approach to these systems are the ones who get the greatest benefit.

Supply chain management (SCM) software is also benefiting from what is known as supplier relationship management (SRM) software, customer relationship management (CRM), and product lifecycle management (PLM) software.

SRM software is a subsystem of SCM software that helps to automate, simplify, and accelerate the procurement-to-pay processes for goods and services.

CRM software originally was a standalone system directed at sales force automation, marketing, and customer service. More recently, it is becoming more integrated with supply chain software such as ERP systems.

PLM software helps companies to collaborate and manage the entire lifecycle of a product efficiently and cost-effectively, from ideation, design, and manufacture, through service and disposal. It is where applications such as computer-aided design (CAD), computer-aided manufacturing (CAM), computer-aided engineering (CAE), and product data management (PDM) come together.

According to Gartner (2013), the supply chain software market was $8.3 billion in 2012, which was a 7.1% increase over the prior year. This particular software market is very fragmented. (The top 20 vendors account for over half of the market, with the largest 2 vendors, SAP and Oracle, having a 38% combined share, and there are literally hundreds of vendors overall.) Driving this large investment has been the need to be more competitive, reduce risk, operate in the global market, and meet various government regulations and industry standards.

SCM systems can be viewed in terms of planning (SCP) and execution (SCE).

In general, SCP applications apply algorithms to predict future requirements of various kinds and help to balance supply and demand.

SCE software applications usually monitor physical movement and status of goods as well as the management of materials and financial information of all participants in the supply chain.

Supply Chain Planning

SCP software vendors address short- to long-term planning and focus on demand, supply, and the balance of demand and supply together usually in the form of a sales & operational planning (S&OP) process described in more detail later:

- **Demand management:** There are three main functions of demand management software, which are 1) predicting demand, 2) using "what-if" analysis to create sales plans, and 3) using what-if analysis to shape demand. Forecasts are typically a rolling 24–36 months. Modern supply chain systems have moved toward a demand "pull"-driven model, so demand management has moved from a purely forecasting tool to one that optimizes and shapes demand to some extent.

- **Supply management:** This area helps meet demand with minimal resources at the lowest cost. Software functionality typically found in this area includes supply network planning or optimization (SNP), production scheduling (sometimes referred to as advanced planning systems or APS), distribution requirements planning (DRP), replenishment, and procurement.

- **Sales & operational planning:** S&OP, as mentioned earlier in this book, facilitates monthly executive planning meetings to tie together sales, operation, and financial plans, along with the related tasks. Input is collected from demand, capacity, and financial plans, culminating in a consolidated sales and operational plan.

Supply Chain Execution

SCE systems primarily include warehouse management software (WMS) and transportation management software (TMS) and feature planning, scheduling, optimizing, tracking, and performance monitoring:

- **Warehouse management systems:** WMS controls the flow of goods through the warehouse and interfaces to the material handling equipment. They also typically include automated processing of inbound and outbound shipments and the storage of goods. Administrative features can include processing of EDI transactions, planning shipments, resource management, and performance tracking.

- **Transportation management systems:** A TMS helps to manage global transportation needs, including air, sea, ground, and carrier shipments. In terms of transportation acquisition, and dispatching, a TMS may also handle the planning, scheduling, and optimizing of shipments. They also provide tracking of vehicles, including exception management, constraints, collaborating with partners, and monitoring of freight.

Administrative features can include cost allocations, freight auditing, and payment and contract management.

- **Enterprise resource systems (ERP):** Some might not include ERP systems as SCM tools, but a great deal of the functionality is supply chain and logistics related. ERP systems are an extension of an MRP system to tie in all internal processes as well as customers and suppliers. It allows for the automation and integration of many business processes, including finance, accounting, human resources, sales and order entry, raw materials, inventory, purchasing, production scheduling, shipping, resource and production planning, and customer relationship management. An ERP shares common databases and business practices and produces information in real time and coordinates business from supplier evaluation to customer invoicing.

E-businesses must also keep track of and process a tremendous amount of information, and therefore have realized that much of the information they need to run an e-business, such as stock levels at various warehouses, cost of parts, and projected shipping dates, can already be found in their ERP system databases. As a result, a significant part of the online efforts of many e-businesses involve adding web access to an existing ERP system.

ERP systems have the potential to reduce transaction costs and increase the speed and accuracy of information, but can also be expensive and time-consuming to install.

Other Supply Chain Technologies

Other categories of software are also often used in the supply chain, including the following:

- **Supply chain event management:** These are software applications that enable companies to track orders across the supply chain in real time between trading partners, providing managers with a clear picture of how their supply chain is performing. The information provided by these systems allows a company to sense and respond to unanticipated changes to planned supply chain operations. This breed of systems conveys information regarding supply chain processes at a specific event level, such as a handoff from one supply chain entity to another, the commitment of a product to an order, the movement of a shipment between two logistics network nodes, or the placement of a product into storage.

- **Business intelligence (BI):** This category is made up of applications, infrastructure, tools, and best practices, providing analysis of information to improve and optimize decisions and performance. BI tools help to sort through the vast amount of data that has become available through the continuing adoption of SCM technologies.

In addition, there are related tools for supply chain collaboration, data synchronization, and spreadsheets and database software.

In fact, many smaller companies today still operate their primary planning functions using spreadsheets and run their day-to-day operations with accounting systems such as Quick-Books and Peachtree rather than spend the resources on a full-blown ERP system.

SCM System Costs and Options

The final cost of supply chain management software can be three to five times the cost of the software license because it also includes planning, implementation, training, customization, interfaces, hardware, and configuration of the software. SCM software vendors also typically charge a 15% to 20% annual fee for maintenance and technical support.

A new alternative to installed software is what has become known alternatively as software-as-a-service (SaaS), on-demand, or cloud supply chain software.

Cloud supply chain systems may reduce or eliminate upfront software acquisition costs by offering subscription fees for web-based applications, allowing you to "pay as you go" because fees are based on usage. In this model, there are typically no installation or mainte-nance costs for the customer.

The major concern of most potential users is security, which may or may not be as big a risk as imagined. Nevertheless, cloud software represents the single highest growth sector in the enterprise software market, and some software vendors are expanding into the cloud by offering some of their SCM modules as SaaS.

According to Gartner (2013), "Software as a service (SaaS) SCM offerings showed above-market growth (13 percent in 2012), while perpetual new licenses experienced slower growth of 3.5 percent, as organizations focused on fast implementation at a lower upfront cost."

Best-in-Class Versus Single Integrated Solution

For more specialized types of supply chain applications such as network optimization and forecasting, choosing a best-in-class solution may be the way to go because supply chain ven-dors with a single integrated solution are limited to larger vendors such as SAP and Oracle. In many cases, companies may select one vendor for SCP and another for SCE.

When licensing best-in-class software, costs may be greater to implement because they require additional interfaces when having multiple vendor relationships. An application known as an enterprise application interface (EAI) system can reduce some of the integra-tion cost.

The benefits of one integrated solution are many, including having a single point of contact, common user interface, and common IT architecture.

Consultants

The three types of supply chain consultants involved in the technology selection and implementation process are as follows:

- **SCM experts or management consultants:** SCM experts help with the planning and modeling
- **Software vendor consultants:** Consultants employed by the software vendor who are application software subject matter experts (SME) and help implement the software
- **IT consultants:** Information technology (IT) consultants who help with infrastructure, interfaces, and custom programming (Erpsearch.com, 2014)

The number and mix of consultants in an SCM software implementation project will vary depending on the size and scope of the project.

Current and Future Trends in Supply Chain Software

Technology is evolving at an ever increasing rate. This was never more true than in the world of supply chain and logistics management. As a result, it is important that we look at some of the short term and emerging trends in supply chain software.

Short-Term Supply Chain Technology Trends

The 2010 annual Gartner supply chain study (Gilmore, 2010 and 2013) looked at supply chain application areas and where companies say they stand in terms of adoption.

The top application area fully implemented (not including ERP systems) were WMSs, which were fully deployed by only 39% of respondents. That was followed by SCP (32%), S&OP (29%), and TMSs (28%).

The top-three obstacles to achieving their company's supply chain goals were forecast accuracy/demand variability (59%), supply chain network complexity (42%), and lack of internal cross-functional collaboration and visibility (39%).

According to the study, the investment priorities when it came to supply chain technology were "improving planning processes," with 20% of respondents, followed by "aligning corporate and supply chain strategies" and "improving supply chain visibility," both at 11%.

Interestingly, in the 2013 study, partially due to continued sluggish economic growth, the focus on using the supply chain to drive business growth was the top priority of companies surveyed, with customer service in the second spot.

The obstacles to reaching supply chain goals, however, were similar to the ones mentioned previously from the 2010 survey.

Emerging Supply Chain Technology Trends

Although the types of software applications in a supply chain probably won't drastically change, the methods for gathering data and using and sharing applications will.

Major areas of innovation most recently include the following:

- **Cloud computing:** As mentioned previously, this method, also referred to as software-as-a-service or SaaS, delivers a single application through the browser to thousands of customers "on demand," thus avoiding costly licensing, implementation, and maintenance costs. It allows companies to focus on their core competencies while allowing a third party to manage technical elements. Salesforce.com is probably the best-known example among enterprise applications, but it is also common for HR and ERP applications as well as some "desktop" applications such as Google Apps.

- **Mobile computing:** Supply chain execution and event management is becoming more mobile with basic visibility and traceability available on smartphones and other mobile devices.

- **Third-party logistics providers (3PLs) providing technology:** Cost is a big driver of this because 3PLs can offer economies of scale, especially for small and mid-size companies.

- **Radio frequency identification (RFID):** An automatic identification method using electronic tags with an embedded microchip and antenna. RFIDs can be utilized in a variety of forms within the supply chain (for example, they can be embedded in between the cardboard layers in a carton or product packaging).

To become more widespread (the 2010 Gartner study showed 51% of the companies surveyed not doing anything with RFID), RFID costs will need to continue to decline to make it more economically feasible. There are also equipment issues that will need to improve, such as reader range, sensitivity, and durability.

An example of the use of RFIDs is with Intel (Harrington, 2007), the global semiconductor manufacturer that needs to know where its product is at all times. They are embarking on a joint effort with DHL, utilizing sensors to monitor the condition of containers as they move around the world. Intel has another project, in Costa Rica, using RFID technology to minimize handheld scanning of inbound and outbound shipments. They have achieved labor savings of 18+% as a result of faster processing and have also eliminated steps in these processes (Harrington, 2007).

Examples of Emerging Technologies in Use Today

Stanley Steemer

A national carpet cleaning franchise founded in 1947, they have automated route operations at two of their branches, with mobile computers with integrated wide-area wireless connectivity, GPS, and a magnetic stripe reader to process credit card payments in real time when service is finished.

Dispatching dynamically is enabled by the GPS and real-time two-way communication. This allowed them to improve efficiency so that they were able to eliminate one full-time dispatcher position at each branch, each of which has significantly reduced paperwork processing time.

Mission Foods

Mission Foods is one of the world's largest producers of tortillas, with its products sold throughout the United States in a direct-store-delivery environment with supermarkets and retailers. They switched from manual invoicing to invoicing with a handheld computer and printing a copy for the customer with a mobile printer. Invoices are sent electronically in real time to Mission headquarters over a wide-area wireless network.

This has eliminated the need for Mission Foods to scan and process thousands of paper invoices (not to mention savings in ink, as a thermal printer will be used). The small wearable printers will save drivers time and fatigue from climbing in and out of the vehicle to print invoices.

The application is a good example of the value of converged wireless.

Lighthouse for the Blind

Lighthouse for the Blind is a nonprofit organization that trains and employs non-sighted workers. They were able to improve their warehouse picking accuracy by 25% with a new speech-recognition system. This system has an audible confirmation of picked items, which allows the blind workers to accurately pick orders.

Social Security Administration

The U.S. Social Security Administration (SSA) has implemented RFID systems in one of its warehouses, enabling them to track inventory and ship more efficiently to branch offices. This resulted in a 39% productivity improvement and a $1 million annual savings as well as a 70% labor savings (Intermec Technologies Corporation, 2007).

There is also emerging supply chain technology being developed now that will have a major impact in the near future, including the following:

- **Multi-enterprise visibility systems:** These are systems providing a comprehensive and timely view of processes, solutions, and metrics across the entire value chain.

When implementing collaborative programs such as VMI (vendor-managed inventory), outsourcing, or JIT (just-in-time), it is important to also implement the infrastructure or processes necessary to manage inventory in this extended supply chain. This type of emerging solution offers a 360-degree view of supply chain events.

- **People-enabling software:** This is technology that empowers people to analyze, find, use, collaborate, and to share data to maximize efficiency and workflow. ERP and other enterprise software solutions help enable and automate business processes, but they only alert users when problems occur; they don't help solve the problems themselves. Technology companies are coming out with productivity tools that enable people to combine unstructured information and business processes with the structured business processes that ERP applications provide. This type of technology platform would enable users to handle multiple alerts to a smartphone; for example, empowering them to put the fires out on the spot by connecting customer, manufacturer/distributor, and supplier systems on a mobile device.

- **Execution-driven planning solutions:** These are tools that utilize data from current executed processes to drive future planning and forecasting. Over the years, it has been common for many companies to have a disconnect between planning and execution. These systems will use information as to the current state of a business to help drive planning decisions for the next planning period in real time. This will enable users to consolidate and aggregate massive amounts of data in meaningful ways by applying machine learning techniques to data-mining algorithms to detect data trends. This will give businesses the chance to respond to problems and take advantage of opportunities much faster than before.

- **Human supply chain technology:** These are solutions that apply supply chain technology to the management of human resources (that is, the labor supply chain) and allow companies to standardize job descriptions, capture spend and labor rates, and improve their hiring practices.

At this point, you should have a fundamental understanding of the supply chain and logistics function as well as the technology used to enable it. We will now take a step back to examine the more strategic concept of supply chain and logistics network design and its impact on cost and efficiency.

PART V

Supply Chain and Logistics Network Design

15

Facility Location Decision

s the saying goes, "In retail, the three most important factors for success are location, location, and location." This is not only true in retail but also for manufacturers because it plays a major role in both the cost and service of the organization.

Unfortunately, it is usually not considered often enough in the case of manufacturers or, in many cases, in enough detail, and can potentially lead to catastrophe.

Location is a strategic, long-term decision in nature that is not easily changed in the short term and applies to raw material sourcing, manufacturing, distribution, and retail.

Strategically, the major goal or priority of the location decision for a manufacturer is to minimize cost, whereas retailers look to maximize revenue where possible.

The Importance of Facility Location When Designing a Supply Chain

To remain competitive in today's global economy, the efficient movement of goods from raw material sites to processing facilities, manufacturers, distributors, retailers, and customers is critical.

Unlike transportation and inventory decisions, location decisions tend to be less flexible because many of the costs are fixed in the short term.

Picking the wrong manufacturing or distribution location can have a long-term impact on the total cost of a product. This decision can be heavily influenced by transportation costs; they can average 3% to 5% of sales, with warehousing costs being 1% to 2% on average, historically.

Because this is a supply chain book, we primarily concentrate on the distribution facility location decision, which is important because it is a key driver of the overall profitability of a firm because it affects both supply chain cost and what the customer experiences directly.

The distribution network can be used to achieve a variety of supply chain objectives discussed earlier in this book, such as a cost- or responsiveness-focused strategy.

For example, Dell computer primarily ships directly to consumers, whereas other PC manufacturers such as Asus sell through retailers. So, although the Dell customer may have to wait several days to get a highly customized PC, retail customers can take their more standardized PC home the day they buy it locally.

Large household and personal-care companies such as Colgate distribute directly to the larger supermarket chains, whereas small retailers have to buy Colgate products from distributors because they buy in smaller quantities.

Supply Chain Network Design Influencers

Like many aspects of the supply chain, there are many tradeoffs. In terms of supply chain network design, there is the major tradeoff of cost versus service.

From the customer perspective, service may be viewed in a variety of ways, including the following:

- **Lead time:** The amount of time for a customer to receive an order.
- **Product variety:** The number of different products offered by a distribution network.
- **Product availability:** The likelihood of a product being in stock when the customer places their order.
- **Customer experience:** This may have many dimensions, including how easy it is for customers to place and receive orders as well as how much the experience is customized.
- **Time to market:** The time it takes to develop new product and bring them to market.
- **Order visibility:** The ability of customers to track orders from time of placement to delivery.
- **Returnability:** How easy it is for a customer to return merchandise and the efficiency of the network to handle these returns.

In general, companies selling to customers who can handle a relatively long response time may require only a few locations, far from the customer. In these cases, companies may concentrate on increasing the capacity of each location.

However, companies that sell to customers who are looking for short response times, and maybe even picking up product with their own vehicles, need to locate facilities close to them. These companies typically have many facilities, each with relatively low capacity.

So, the tradeoff here is that faster response times required to meet customer demand increases the number of facilities required in the network, and, conversely, a decrease in the response time increases the customer's desire and the number of facilities required in the network (see Figure 15.1).

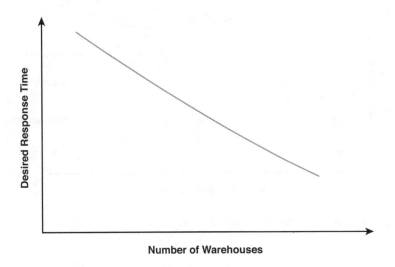

Figure 15.1 Relationship between number of warehouses and response time

Changing the distribution network design affects other supply chain costs (see Figure 15.2), such as the following:

- **Inventories:** The more locations, the harder to accurately forecast demand because there are smaller and smaller demand groupings, making the target smaller and harder to hit. As a result, safety stock requirements go up almost exponentially (that is, the *square root* rule discussed in Chapter 8, "Warehouse Management and Operations").

- **Transportation:** Ideally, we want a *long in and short out* to gain economies in transportation on the inbound end (that is, full truckloads versus less-than-truckload [LTL]), but this, of course, can reach a point of diminishing returns because it will increase inventory and warehouse operating costs.

- **Facilities and handling (that is, warehouse operations):** Certain economies of scale are gained by operating fewer warehouses, whether company owned or outsourced by consolidating volume. This can result in lower unit handling and storage costs, with fewer facilities, and must be analyzed thoroughly. However, the use of local public distribution centers (DCs) may enable a company to have their products combined with other companies' products to gain some local transportation savings.

There is also the fact that as the number of distribution facilities increases, the amount of information to manage increases. This can be somewhat mitigated by having efficient and integrated information technology systems.

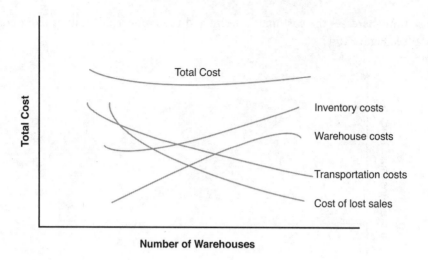

Figure 15.2 The number of warehouses and the impact on cost

Types of Distribution Networks

Keeping the previously mentioned influences in mind, a company can distribute its products in a variety of ways, as discussed throughout this section.

Manufacturer Storage with Direct Shipping

In this type of distribution network design (see Figure 15.3), product is shipped directly from the manufacturer to the end customer, bypassing the retailer or seller (who takes the order and initiates the delivery request). This is also referred to as *drop shipping*, where product is delivered directly from the manufacturer to the customer. This tends to work for a large variety of low-demand, high-value items where customers are willing to wait for delivery and accept several partial shipments.

Impact on Costs

In general, this type of network has lower costs because of aggregation, which works best with low-demand and high-value items. Transportation costs are greater because of increased distance and individual item shipping. Facility costs are lower due to this aggregation of demand, and there may be some saving on handling costs if the manufacturer can directly ship these small orders from the production line. However, this type of design requires a fairly large investment in information infrastructure because the manufacturer and retailer need to be tightly integrated.

Figure 15.3 Manufacturer storage with direct shipping (drop shipping)

Impact on Service

In terms of service, this type of distribution network design requires fairly long response times of 1 to 2 weeks because of increased distance and the two stages for order processing. The response time may vary by product, which may complicate receiving. Product variety and availability are relatively easy to provide due to aggregation at the manufacturer. Home delivery may result in high customer satisfaction, but this can be negatively affected if orders from multiple manufacturers are sent as partial shipments. This type of network can help to get products to market fast, with the product available as soon as the first unit is produced. However, customer visibility and product returnability may be more difficult and expensive.

Manufacturer Storage with Direct Shipping and In-Transit Merge

In-transit merge by a carrier combines pieces of an order coming from different locations so that the customer gets a single delivery (see Figure 15.4).

For example, when a customer orders a PC from Hewlett Packard (HP) along with a Samsung monitor, the package carrier picks up the PC from the Samsung factory and the monitor from the HP factory; it then merges the two together at a hub before making a single delivery to the customer.

This type of network works best for low- to medium-demand, high-value items that a retailer is sourcing from a relatively low number of manufacturers.

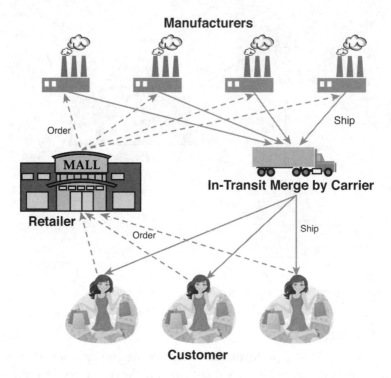

Manufacturers

Order

Ship

MALL

Retailer

In-Transit Merge by Carrier

Order

Ship

Customer

Figure 15.4 Manufacturer storage with direct shipping and in-transit merge

Impact on Costs

The inventory costs associated with this type of distribution network are similar to drop shipping. However, handling and information investment costs may be higher than drop shipping, but transportation costs and receiving costs at the customer are somewhat lower than drop shipping. As a result of more coordination required to combine shipments, the information investment is somewhat higher than for drop shipping.

Impact on Service

The impacts on service, such as response time, variety, availability, visibility, and returnability, are all similar to drop shipping. However, the customer experience may be better than drop shipping because a single order has to be received rather than multiple orders.

Distributor Storage with Carrier Delivery

When using the distributor storage with carrier delivery option (see Figure 15.5), inventory is not held by manufacturers at the factories but instead is held by distributors/retailers in intermediate warehouses, and package carriers are used to transport products from the intermediate location to the final customer. This works well for medium- to fast-moving items. It also makes sense when customers want delivery faster than is offered by manufacturer storage but do not need it immediately.

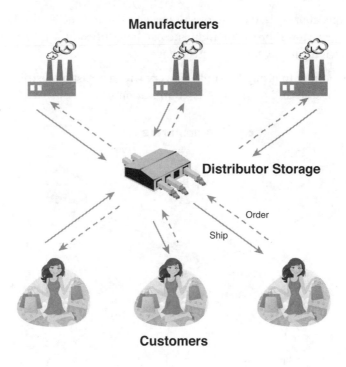

Figure 15.5 Distributor storage with carrier delivery

Impact on Costs

In this type of configuration, inventory and warehouse operations costs are higher than manufacturer storage with direct shipping and in-transit merge. Transportation costs are lower than manufacturer storage, with a simpler information infrastructure required when compared to manufacturer storage.

Impact on Service

Distributor storage with carrier delivery typically has faster response time than manufacturer storage with drop ship, but offers less product variety and higher product availability costs. The customer experience, order visibility, and product returns are better than manufacturer storage with drop shipping.

Distributor Storage with Last-Mile Delivery

Last-mile delivery refers to the distributor/retailer delivering the product to the customer's home instead of using a package carrier (see Figure 15.6).

In areas with high labor costs, distributor storage with last-mile delivery is hard to justify on the basis of efficiency or improved margin and can be justified only if there is large enough demand that is willing to pay for this convenience.

It is always a good idea to group last-mile delivery with an existing distribution network to gain economies of scale and to improve asset utilization.

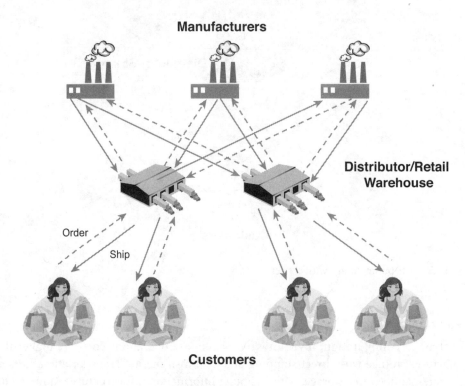

Figure 15.6 Distributor storage with last-mile delivery

Impact on Costs

Distributor storage with last-mile delivery inventory costs more than distributor storage with package carrier delivery. Warehouse operations costs are greater than manufacturer storage and distributor storage but lower than the costs of a retail chain. The transportation costs are greater than any other distribution network option. Information costs are similar to distributor storage with package carrier delivery.

Impact on Service

Service response times are very quick and in some cases can be same-day to next-day delivery, with a very good customer experience, particularly for bulky items. Product variety is less than distributor storage with package carrier delivery but greater than retail stores, with availability being more expensive to provide than any other option except retail stores. There is less of an issue of order traceability than manufacturer storage or distributor storage with package carrier delivery, and returnability is easier to implement than other options, except perhaps a retail network.

Manufacturer or Distributor Storage with Customer Pickup

Manufacturer or distributor storage with customer pickup (see Figure 15.7), involves inventory being stored at the manufacturer or distributor warehouse but customers placing their orders online or on the phone and then having to travel to designated pickup points to collect their merchandise. Orders are shipped from the storage site to the pickup points as needed. Such a network is likely to be most effective if existing locations, such as coffee shops, convenience stores, or grocery stores, are used as pickup sites, because this type of network improves the economies from existing infrastructure.

Impact on Costs

Manufacturer or distributor storage with customer pickup is similar to the other distribution configurations in terms of inventory costs. Transportation costs are on the low side because there is not a great use of package carriers, especially if using an existing delivery network (plus customers pickup themselves). Warehouse operations costs can be high if new facilities have to be built, and lower if existing facilities are in place (and handling costs at the pickup site can be fairly high). Information costs to provide infrastructure in this option can be very high as well.

Impact on Service

Response times are similar to package carrier delivery with manufacturer or distributor storage with same-day delivery possible when items are already stored locally at pickup sites. Product variety and availability are similar to other manufacturer or distributor storage options. Order visibility is extremely important and can be greatly aided with the help of technology. Product returns are somewhat easier because pickup locations can typically process returns. The customer experience may be lower than the other options due to the lack of home delivery, but in densely populated areas, the loss of convenience may be small.

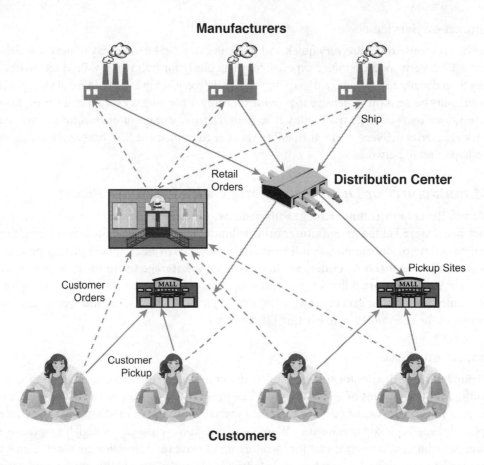

Figure 15.7 Manufacturer or distributor storage with customer pickup

Retailer Storage with Customer Pickup

Retailer storage with customer pickup is, of course, the most common form of distribution network, where inventory is stored locally at retail stores. Inventory can be supplied to the stores from the retailer warehouse in the case of a chain, a distributor/wholesaler warehouse, or even direct to the store from the manufacturer (factory or warehouse) for larger retailers. Customers walk into the retail store or place an order online or by phone and pick it up at the retail store. This option is best for faster-moving items or items for which customers want quick response.

Impact on Costs

Retailer storage with customer pickup has the highest inventory and warehouse operations costs and lowest transportation costs of all the options. There may be an increase in handling cost at the pickup site for online and phone orders, which may also require some investment in infrastructure.

Impact on Service

Response times are the quickest of the options, because same-day pickup is possible for items stored locally at the retail location. Product variety, although great, is lower than the other options, and availability is more expensive to provide than all other options. The customer experience may be considered positive or negative based on how shopping is viewed by customer. Order visibility really only applies for online and phone orders, and returnability is easier than other options, given that retail locations can handle returns.

Impact of E-Business on the Distribution Network

Operating via an e-business can have both cost and service impacts on the distribution network.

Cost Impacts

An e-business can reduce its inventory levels and costs by improving supply chain coordination to better match supply and demand. Also, when customers are willing to wait for delivery of online orders, e-business can enable a firm to aggregate inventories remotely from customers.

An e-business can reduce network facility costs (that is, costs related to the number of facilities in a network) by centralizing operations, thereby decreasing the number of facilities required. In addition, they can also lower operating costs by allowing customer participation in selection and order placement.

In general, aggregating inventories, often the case with e-businesses, increases outbound transportation costs relative to inbound transportation costs. If a firm's product is in a form that can be downloaded, the Internet will allow it to save on the cost and time for delivery (for example, downloadable music and software).

An e-business can share demand information throughout its supply chain to improve visibility more readily than a bricks-and-mortar business. This can help reduce overall supply chain costs and better match supply and demand.

Service Impacts on the Distribution Network

Response time to customers for products that can be downloaded, such as a mutual fund prospectus or music, are generally very fast. However, e-businesses without a physical retail outlet selling physical products take longer to fulfill a customer request than a retail store because of the shipping time involved.

An e-business can offer a much larger selection of products than a bricks-and-mortar store because a retail store requires a large location with a large amount of inventory to offer the same variety.

Because e-businesses have greater speed with which information on customer demand is shared throughout the supply chain, they tend to have more timely and accurate forecasts, resulting in a better match between supply and demand.

An e-business affects customer experience in terms of access, customization, and convenience and allows access to customers who may not be able to place orders during regular business hours and access to customers who are located far away. The Internet is ideal to organizations that focus on mass customization to help customers select a product that more closely suits their needs. On top of that, customers have the benefit of not having to leave home or work to make a purchase.

The Internet makes it possible to provide visibility of order status, especially important for an online order because there is really no way to match a customer shopping in person.

The proportion of returns for online orders is typically much higher (and expensive because they are usually shipped from a central location) because customers cannot touch and feel the product before their purchase arrives. To counteract this effect, there is now the fairly common practice of *showrooming*, where customers examine merchandise in a traditional brick-and-mortar retail store and then buy it online, sometimes at a lower price.

Having an e-business also allows manufacturers and other members of the supply chain that do not typically have direct contact with customers in traditional channels to increase revenues by skipping intermediaries and selling directly to customers in some cases. However, great caution must be taken by the manufacturers so as not to directly compete with its distribution and retail customers. So, in many cases, this may be used as a way to run out discontinued, excess, or even to sell new/different items.

It is also fairly easy for an e-business to adjust prices at the click of a button on their website. In this way, they can maximize revenues by setting prices based on current inventories and demand. (For example, Amazon.com often adjusts book prices, up and down, based on rankings and other demand criteria.) Not to mention that e-businesses can enhance revenues by speeding up collection versus the 60- to 90-day payment terms found in traditional channels for manufacturers and distributors.

Finally, e-businesses can introduce new products more quickly than organizations that use traditional physical channels because they can rapidly introduce a new product by making it available on the website, versus a traditional manufacturer who faces a lag to fill their distribution channel pipelines.

Location Decisions

There are a set of decisions that an organization must go through in order to make the best facility location decisions for their supply chain(s). They range from strategic decisions to ensure the locations fit an organizations competitive strategy to more tactical one that involve identifying specific geographic locations and sites.

Strategic Considerations

The first thing you need to think about in terms of the location decision is your particular organization's competitive strategy, which will drive the design of your supply chain network. For example, a strategy of cost leadership will result in a very different supply chain network than one that is primarily based on responsiveness or product differentiation.

The objective of location strategy is to maximize the benefit of location to the firm. In the case of a manufacturer or distributor, you want to focus on cost minimization while meeting service goals. For retail, it is more about maximizing revenue.

Other things to consider at a more strategic level are identifying your key competitors in each target market as well as your capital constraints.

As you grow, you need to consider whether to reuse existing facilities, build new facilities, or partner with other companies (or all of these).

Selecting a New Facility

When selecting a new facility location, it is important to go through some general steps, as follows (see Figure 15.8):

1. Identify the important location factors and categorize them as key success factors (KSFs) or secondary factors.
2. Consider alternative countries, and then narrow them down to alternative regions/ communities and then finally to specific sites.
3. Collect data on the location alternatives.
4. Analyze the data collected, starting with quantitative factors.
5. Merge the qualitative and quantitative factors relevant to each site into the evaluation.

Steps 1 and 2 in the location decision can also be accomplished by starting at a predetermined country level, working your way down to the local or site decision.

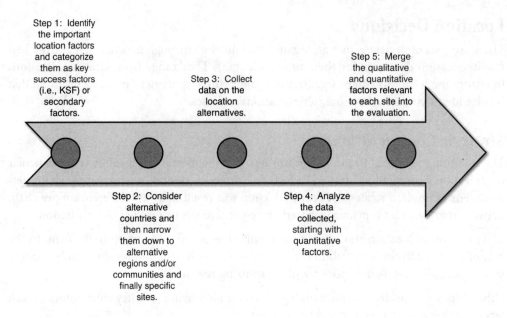

Step 1: Identify the important location factors and categorize them as key success factors (i.e., KSF) or secondary factors.

Step 3: Collect data on the location alternatives.

Step 5: Merge the qualitative and quantitative factors relevant to each site into the evaluation.

Step 2: Consider alternative countries and then narrow them down to alternative regions and/or communities and finally specific sites.

Step 4: Analyze the data collected, starting with quantitative factors.

Figure 15.8 Steps in a new facility location decision

Location Decision Hierarchy

An organization's location decision, whether in goods or service, may be at the national (within a continent), regional (within a country), or a local selection level (Heizer & Render, 2013). This will vary depending on where you are in the location decision for your business. Each decision may be different.

The criteria (KSFs) used to determine the best location will vary based on the levels previously mentioned.

Country Decision

The country decision, especially in today's global economy, may involve a wide range of factors to consider, including the following:

- Tariffs and tax incentives
- Infrastructure factors
- Exchange rate fluctuations and currency risk
- Demand and supply risk
- Competitive environment
- Political risks, government rules, attitudes, and incentives

- Cultural and economic issues
- Location and demand of markets
- Labor talent, attitudes, productivity, costs
- Availability of supplies, communications, and energy

Regional Decision

Once the country decision has been made, an organization must determine which region is best for its new location.

Factors to consider include the following:

- Corporate desires
- Attractiveness of region
- Labor availability and costs
- Costs and availability of utilities
- Environmental regulations
- Government incentives and fiscal policies
- Proximity to raw materials and customers
- Land/construction costs

We've all seen states competing against each other to land a new large manufacturing facility, offering all kinds of incentives such as tax reductions/postponement, rebates, and so on. So, some of these listed factors may be *artificially* modified to influence this decision.

Local Decision

After a company has settled on a location within a relatively small metropolitan area, it must consider which specific site location is best for its needs. Factors in the site decision process include the following:

- Site size and cost
- Air, rail, highway, and waterway systems
- Zoning restrictions
- Proximity of services/supplies needed
- Environmental impact issues

The relative importance of these factors will vary depending on the size and industry of your organization. For example, some towns are zoned only for light industry, or your company might need easy access to major highways and airports for inbound and outbound transportation needs.

Certain types of factors are more important in the goods versus services location decision.

Dominant Factors in Manufacturing

The factors most important to the manufacturing location decision include the following:

- Favorable labor climate
- Proximity to markets
- Impact on environment
- Quality of life
- Proximity to suppliers and resources
- Proximity to the parent company's facilities
- Utilities, taxes, and real estate costs

Dominant Factors in Services

The major factors to consider in the service location decision include the following:

- Impact of location on sales and customer satisfaction
- Proximity to customers
- Transportation costs and proximity to markets
- Location of competitors
- Site-specific factors

Location Techniques

Steps 3 through 5 in the new facility selection decision process, as outlined in Figure 15.8, involve gathering quantitative data and then combining qualitative and quantitative data to determine the specific site location. You can use a variety of techniques to accomplish this, as discussed throughout this section.

Location Cost-Volume Analysis

A relatively simple location decision tool is known as *cost-volume* (CV) analysis. It can be represented either mathematically or graphically and is basically the same as a traditional breakeven analysis or crossover chart used in a variety of situations in business. The general idea is based on a future volume forecast with fixed and variable costs known for each location, to determine which candidate location can be justified by the predicted throughput volume.

CV involves three steps:

1. For each location alternative, determine the fixed and variable costs.
2. For all locations, plot the total cost lines on the same graph.
3. Use the lines to determine which alternatives will have the highest and lowest total costs for expected levels of output.

In addition, you want to keep four assumptions in mind when using this method (making it rather simplistic, but often giving a *close enough* answer to determine the best facility from a list of candidates):

- Fixed costs are constant.
- Variable costs are linear.
- Required level of output can be closely estimated.
- Only one product is involved.

The location with the lowest total cost [(Fixed cost + (Variable cost × Volume)] is the location selected.

Table 15.1 shows an example of location cost-volume analysis. It assumes a selling price of $100/unit with an expected volume of 5,000 units.

Table 15.1   Location Cost-Volume Example Data

City	Fixed Costs	Variable Costs	Total Costs
Hong Kong	$700,000	$50	$750,000
Singapore	$400,000	$75	$725,000
Ho Chi Minh City	$200,000	$100	$700,000

The *crossover point* from Ho Chi Minh City to Singapore (that is, justifying Singapore) would be:

$$200,000 + 100(x) = 400,000 + 75(x).$$

$$25(x) = 200,000$$

$$(x) = 8,000 \text{ units}$$

The *crossover point* from Singapore to Hong Kong would be:

$$400,000 + 75(x) = 700,000 + 50(x)$$

$$25(x) = 300,000$$

$$(x) = 12,000 \text{ units}$$

Graphically, this can be expressed as shown in Figure 15.9.

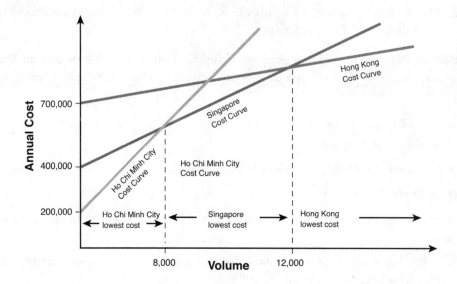

Figure 15.9 Location cost-volume analysis example

Of course, like with many of these tools, this is dependent upon a volume forecast. So, it is wise to do *what-if* analysis with different levels of cost and volumes to see the impact on the decision.

Weighted Factor Rating Method

The weighted factor rating method compares a number of locations using both quantitative and qualitative criteria.

The first step is to identify the factors that are key to the success of the facility at the location. Then you assign a weight as to the importance of each factor (the total of which must sum to 100%).

The next step is to determine a score for each factor. Usually, a multifunctional team is involved in this process. At this point, a lot of data has been gathered on the potential sites from research, visits to the sites, and even presentations by the communities, states, or countries extolling the virtues (and incentives) to choose them.

You then multiply the factor score by the weight and sum the weighted scores. The location with the highest total weighted score is the recommended location.

This is often a useful tool in *whittling down* a list of candidates and potentially selecting one. Often, other factors may come into play, some logical (for example, extra tax incentives) and some not so logical. (The owner of the company has a vacation home in the state or country that came in second, for example.)

The example in Table 15.2 illustrates the weighted factor rating method.

Table 15.2 Weighted Factor Rating Method Example

SCORES (weight *100 points maximum)				
Factor	Weight	Country A	Country B	Country C
Stability of government	.20	.20 * 90	.20 * 75	.20 * 80
Labor cost	.30	.30 * 50	.30 * 80	.30 * 85
Education and health	.20	.20 * 70	.20 * 50	.20 * 60
Tax structure	.30	.30 * 70	.30 * 90	.30 * 80
Totals:	1.00	68.0	76.0	77.5

Center of Gravity Method

The center of gravity method of location analysis is a technique used in determining the location of a facility that will either reduce travel time or lower shipping costs. Specifically, it determines the location of a DC that minimizes distribution costs while considering the location of markets, volume of goods shipped to those markets, and shipping cost (or distance).

In this model, distribution costs are seen as a linear function of the distance and quantity shipped. The center of gravity method uses a visual map and a coordinate system. The x and y coordinate points are placed on an XY chart with an arbitrary starting point, but located in similar positions for each location because they would be on a map and are treated as a set of numeric values when calculating averages. If the quantities shipped to each location are the same (rarely the case), the center of gravity is found by taking the averages of the x and y coordinates; if the quantities shipped to each location are different (more typical), a weighted average must be applied (the weights being the quantities shipped).

Years ago, I'm told, this method was actually demonstrated using strings and weights and thus the name *center of gravity*.

This method is best described graphically using the following example.

If a retailer has four stores located in the following cities with the listed demand, what would be the best location, or *center of gravity*, for a DC that minimizes shipping costs (see Table 15.3 for demand by location)?

Table 15.3   Center of Gravity Example

Store Location	Truckloads Shipped to Store/Month
Chicago	5,000
Cleveland	4,000
Cincinnati	3,000
Indianapolis	2,500

We will need to first need to determine the *x* and *y* coordinates for each of the store locations using an XY chart (see Figure 15.10).

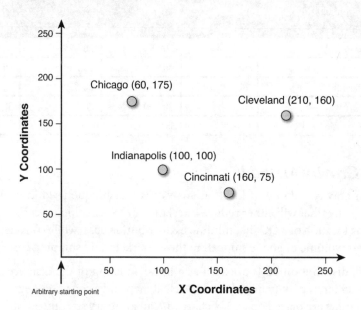

Figure 15.10 XY chart for center of gravity method example

Next, we will use the *x* and *y* coordinates as well as the retail location demand to calculate the center of gravity for the DC location.

The calculation is [(Sum of shipping volume * *x* or *y* coordinate) / Total system shipping volume] and is the same to calculate the center of gravity for the *x* and *y* coordinates (of course, using the appropriate coordinates each time).

So, in this example, when we solve for *x* and *y*, we find the following:

x coordinate = (60 * 5,000) + (210 * 4,000) + (160 * 3,000) + (100 * 2,500) / (5,000 + 4,000 + 3,000 + 2,500) = **129.0**

y coordinate = (175 * 5,000) + (160 * 4,000) + (75 * 3,000) + (100 * 2,500) / (5,000 + 4,000 + 3,000 + 2,500) = **137.2**

The center of gravity solution can then be placed on the XY chart (see Figure 15.11). The actual physical location can be approximated by placing a map over the chart (or using your own geographic knowledge to approximate the location) to get you down to the site location decision.

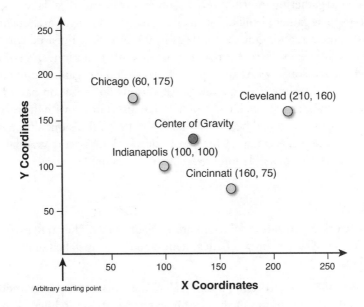

Figure 15.11 Solved XY chart for center of gravity method example

The Transportation Problem Model

The transportation problem is a special class of the linear programming model that deals with the distribution of goods from several points of supply (that is, sources) to a number of points of demand (that is, destinations).

Usually, we are given the capacity of goods at each source and the requirements at each destination as well as a variety of costs for materials, manufacturing, transportation, and warehousing.

Typically, the objective of the transportation problem with linear programming is to minimize total transportation and production costs while maintaining specified inventory and service-level targets.

Fairly complex software solutions are available today that, using this or other algorithms, will provide a solution that can not only be the optimal solution in terms of the number and location of distribution facilities in your network but also, through the use of what-if analysis, test multiple scenarios looking at changes in demand, transportation, and material costs, for example.

In many cases, these types of studies are performed by external consultants for fees of up to $100,000 for medium- to large-size organizations. The payoff, however, can be in the tens of millions, so it is well worth the cost of doing one of these studies every 3 to 5 years.

Getting started and gathering information can sometimes be the most daunting task because much data needs to be gathered, including material, manufacturing, warehousing, and transportation costs and demand. Once the data is in the model, it must be *validated* against previous periods to make sure that it matches actual costs and service. Next, a *baseline* or *base case* model is developed with future projections with the organization's current supply chain network. This baseline is then run against various optimization parameters such as lowest cost, fewest DCs, and so on. Finally, rather than just running with the revised network, various what-if scenarios should be run to make sure that a decision, although optimal, is also reasonable if conditions change. This may indicate taking a more gradual wait-and-see approach to distribution consolidation or expansion for example.

Technology

A variety of network optimization solutions are available today. They range from standalone systems to modules of larger supply chain systems and can be installed or on-demand cloud software systems.

JDA, JD Edwards (Oracle), SAP, and Logility all have network optimization modules that are integrated with their other supply chain planning and execution modules.

Other systems, such as IBM ILOG LogicNet Plus XE and Logistix Solutions (see Figure 15.12), offer a standalone system (on demand in the case of Logistics Solutions) for lower upfront costs, but because they are not integrated with other supply chain planning and execution modules, they are perhaps a bit more data intensive.

Careers

Although there is no specific career path for supply chain network analysis, in many cases, as mentioned, consultants perform this type of study. There are also software vendors who both license and train users at organizations to use it for their company. In many cases, this responsibility falls on the shoulders of an internal logistics analyst to run the numbers and do the analysis with recommendations. The actual decisions are usually made by senior management.

After the location decision has been made, some effort both upfront and on an ongoing basis must be given to having a safe and efficient layout, which we cover in the next chapter.

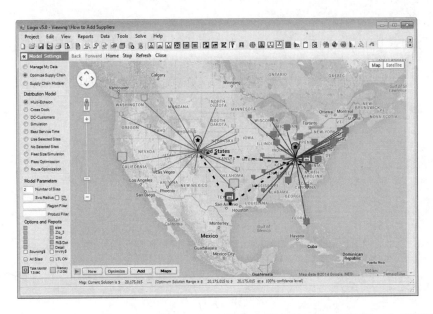

Figure 15.12 Logix screenshot (copyright Logistix Solutions, LLC, Map Data 2014 Google INEGI)

16

Facility Layout Decision

After selecting a facility's location, the next major decision is to design the best physical layout for the facility. The available space needs to be assessed with workstations, equipment, and storage; and other amenities need to be arranged. The goal is to create the most efficient workflow necessary to produce its goods or services at the highest level of quality with the lowest possible cost.

Layout planning is organizationally important not only for efficient operations but also for other functions that are impacted as well, such as marketing, which is affected by layout when clients come to the site, human resources because layout impacts people, and finance because layout changes can be costly endeavors.

The layout decision can determine how efficient a facility is. In the case of the supply chain, this is primarily focused in the warehousing function, in both the warehouse itself and various other areas, including office and maintenance areas.

Layout should be considered in a variety of situations, including when a new facility is being constructed, when there is a significant change in demand or throughput volume, when a new good or service is introduced to the customer benefit package, or then different processes, equipment, or technology are installed.

The focus of layout improvements is to minimize delays in materials handling and customer movement, maintain flexibility, use labor and space effectively, promote high employee morale and customer satisfaction, provide for good housekeeping and maintenance, and enhance sales as appropriate in manufacturing and service facilities.

Types of Layouts

Managers can choose from five primary types of workflow layouts:

- **Product layout:** Production line (for example, an automobile assembly plant)
- **Process layout:** Arranged in departments (for example, hospitals, printer)
- **Hybrid layout:** A combination of both product and process layouts

- **Fixed-position layout:** Building a large item (for example, airplane, cruise ship)
- **Cellular layout:** Reorganizes people and machines into groups to focus on single products or product groups

We will now discuss each in some detail.

Product Layouts

Product arrangements are based on the sequence of operations that are performed during the manufacturing of a fairly standardized good or delivery of a service. Typically, workstations and equipment are located along the line of production, as with an assembly line, for example. Batches of semi-finished (that is, work in process) goods are passed to the next station in a production line.

Some examples of this type of layout include the following: winemaking industry, credit card processing, submarine sandwich shops, paper manufacturers, insurance policy processing, and automobile assembly lines.

Advantages of product layouts include lower work-in-process inventories, shorter processing times, less material handling, lower labor skills, and relatively simple planning and control systems.

Disadvantages include that a breakdown at one workstation can cause the entire process to shut down or a change in product design and the introduction of new products may require major changes in the layout, resulting in limited flexibility.

Process Layouts

Process layouts usually have a functional grouping of equipment or activities that perform similar work.

Examples of process layout including the following: legal offices, print shops, footwear manufacturing, and hospitals.

Advantages of process layouts may include a lower investment in equipment and that the diversity of jobs can lead to increased worker satisfaction.

Disadvantages may include high movement and transportation costs, more complex scheduling and control systems, longer total processing time, higher in-process inventory or waiting time, and higher worker skill requirements.

Warehouse (Process) Layout Considerations

The objective in warehouse (and really all process type) layout is to optimize the tradeoffs between handling costs and costs that are associated with warehouse space while at the same time minimizing damage and spoilage to the product.

When considering warehouse layouts, it is most effective to arrange work centers or functional process areas so as to minimize the total costs of material handling between departments or work centers.

The basic cost elements involved in this *load distance* (LD) minimization calculation are as follows:

- The number of loads (or people) moving between centers
- The distance loads (or people) moved between centers

Once estimating these elements for all possible combinations, a total *load x distance* is calculated for the current state. You can then look for improved layout arrangements that reduce the total *load x distance* by evaluating various number and distance (can use cost, which will vary by distance traveled) of load combinations between the elements. This can be estimated manually in a simple spreadsheet, or for more optimal results, you can use packaged software such as Factory Flow, Proplanner, and CRAFT.

Maximizing Density

By maximizing the total *cube* or space of a warehouse, you are able to better utilize its full volume while maintaining low material handling costs.

Material handling costs include all costs associated with a transaction, such as incoming transport, storage, finding and moving material, outgoing transport, equipment, people, material, supervision, insurance, and depreciation.

Warehouse density also tends to vary inversely with the number of different items stored. Although this might sound counterintuitive, you must realize that each item or stock keeping unit (SKU) will have its own set of dimensions. So, if you carried one single item in a warehouse, you would be able to use almost every inch of storage space. However, most warehouses have hundreds if not thousands of items, all with different dimensions, making it more difficult to maximize the use of storage space.

That's where the concept of *random stocking* can be used, allowing for the more efficient use of warehouse space. Random stocking can be greatly aided by the use of a warehouse management system (WMS; described in Chapter 8, "Warehouse Management and Operations").

Key tasks in random stocking include maintaining a list of open locations, keeping accurate inventory records, the sequencing of items to minimize travel and picking time, combining of picking orders, and assigning classes of items to particular areas.

Minimizing Travel Time to Maximize Warehouse Efficiency

The concept of *velocity slotting* helps to minimize travel time, which is critical to productivity in a warehouse, as it tries to locate at least some of the faster moving *A* type items closer to the shipping area and slower moving items further away and higher up.

The use of automated storage and retrieval systems (ASRS) can significantly improve warehouse productivity.

Dock location is also a key design element. The primary decision is where to locate each department relative to the dock. It is important to organize departments so as to minimize travel time (can be thought of as a *load x distance* total for measurement purposes).

The usage of *cross-docking*, discussed in Chapter 8, modifies the traditional warehouse layouts. Cross-dock facilities tend to have more docks, less storage space, and less order picking because materials are moved directly from receiving to shipping and are not placed in storage in the warehouse.

Cross-docking requires tight scheduling and accurate shipments; barcode or radio-frequency identification (RFID) is used for advanced shipment notification as materials are unloaded. They also typically require automatic identification systems (AISs) and information systems such as a warehouse management system (WMS).

The location of value-added activities performed at the warehouse, which can enable low-cost and rapid-response strategies, must be factored into the layout decision. These can include the assembly of components, repairs, and customized labeling and packaging, among other activities.

Office (Process) Layout Considerations

When we look at an office (process) layout, we consider the grouping of workers, their equipment, and the space required to provide comfort, safety, and flow of information.

One way to evaluate office layout is by using a *relationship matrix* (see Figure 16.1) in which you can look at the relative importance to various people and functions and, based on the results, revise the layout.

The movement of information is a large factor, but it is constantly changing because of frequent technological advances. In fact, certain advances, like the use of electronic documents, may make the movement of information a nonfactor.

Much of the American workforce works in an office environment, including those carrying out many supply chain and logistics administrative functions. In this environment, human interaction and communication are the primary factors in designing office layouts.

When considering layout in an office, you need to account for both the physical environment and psychological needs of the organization.

One key layout tradeoff is between proximity and privacy. Some companies have gone to a more open-office concept, with no walls and with sound-absorbing ceiling panels. Open-concept offices promote understanding and trust, but may not be appropriate depending on job functions and privacy issues.

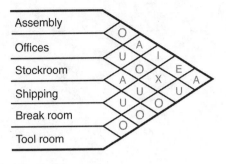

Relationship:

A – Absolutely necessary
E – Especially important
I – Important
O – Ordinary or OK
U – Unimportant
X – Undesirable

Figure 16.1 Office relationship matrix

Flexible layouts can incorporate what is known as *office landscaping* to help solve the privacy issue in open-office environments. Office landscape involves furniture and desk placement, usually in open-plan office settings. It often also involves the selection and placement of plants, the creative use of natural light, and the use of artwork to create ambiance.

Hybrid Layouts

Hybrid or combination layouts combine elements of both product and process layouts. They tend to keep some of the efficiencies of product layouts while maintaining some of the flexibility of process layouts.

For example, a business may have a process layout for the majority of its process but also have an assembly line in one particular area. Alternatively, a firm may utilize a fixed-position layout (described later) for the assembly of its final product, but use assembly lines to produce the components and subassemblies that make up the final product (for example, a cruise ship).

Cellular (or Work Cell) Layouts

This type of layout design is not usually established according to functional characteristics of equipment, but instead by creating self-contained groups of equipment (called *work cells*; see Figure 16.2) and dedicated operators needed for producing a particular family of goods or services (by *family*, we mean one or more items or services that have mostly the same steps or processes).

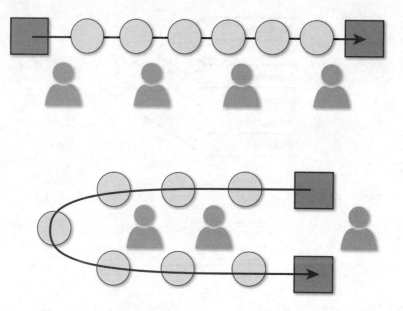

Figure 16.2 Assembly line (top) versus work cell (bottom) layout

The concept of cellular manufacturing or group technology classifies parts into families so that efficient mass-production types of layouts can be designed for the families of goods or services.

Cellular layouts are used to centralize people expertise and equipment capability and are usually laid out in a horseshoe shape rather than as a long assembly line. This is to gain better flow, improve use of equipment, maintain smaller batch size (that is, *make one, pass one*), and require fewer more cross-trained operators.

Work cells, although common in manufacturing, can be applied strategically in offices (for example, processing orders) and warehouses (for example, kitting). Additional examples might include group legal or medical specialties.

Benefits and advantages of work cells include the following:

- Reduced work in process, raw material, and finished goods inventory.
- Less use of floor space.
- Reduced direct labor cost.
- Heightened sense of employee participation as a result of the high level of training, flexibility, and empowerment of employees, which results in improved morale and increased productivity.
- Increased equipment and machinery utilization.

- Because they are essentially self-contained, they have their own equipment and resources.
- Tests such as a poka-yoke, which is a foolproof testing device, are commonly used at stations in work cell to improve quality.

Fixed-Position Layout

A fixed-position layout consolidates resources necessary to manufacture a good or deliver a service, such as people, materials, and equipment, at one physical location. Because the fixed-position layout is typically used with project types of processes, where the product that is too large or too heavy to move, required resources must be portable so that they can be taken to the job for on-the-spot performance.

Production of large items such as heavy machine tools, airplanes, buildings, locomotives, and ships is usually accomplished in a fixed-position layout.

Service-providing firms often use fixed-position layouts, such as major hardware and software installations, sporting events, and concerts.

Due to the nature of the product, the user has little choice in the use of a fixed-position layout.

Disadvantages can include those related to limited space on the site resulting in a work area being crowded, which can also cause material handling problems and administration difficulties, because the span of control can be narrow, making coordination unwieldy.

Facility Design in Service Organizations

Service organizations can use product, process, cellular, or fixed-position layouts to organize different types of work.

Process Layout Examples

There are many examples of process layout in service organization. They can be seen in libraries, which place reference materials, serials, and microfilms into separate areas; hospitals, which group services by function such as maternity, emergency room, surgery, and x-ray; and insurance companies, which have office layouts in which claims, underwriting, and filing are individual departments.

Product Layout Examples

Service organizations that deliver very standardized services tend to use product layouts. For example, the kitchen at a fast-food restaurant that has both dine-in and delivery will tend to be arranged in an assembly line type of layout, with some pre-prepared items such as cooked hamburgers, sliced tomatoes, and so on for easier assembly.

Some eyeglass chains use both process and product layouts; the customer contact area may be arranged in a process layout, but the lab area, where lenses are manufactured, may be in a product layout.

Designing and Improving Product Layouts

There are some fairly common techniques used for the design and improvement of product layouts, which we will now describe in some detail.

Assembly Line Design and Balancing

An assembly line is a product layout dedicated to combining the components of a good or service. Typically, parts are added as the semi-finished item to be assembled moves from workstation to workstation. Parts are added in sequence until the final assembly is produced.

Assembly line balancing is a procedure where tasks along the assembly line are assigned to an individual workstation so that each has roughly the same amount of work.

When designing product layouts in this type of situation, you need to consider the sequence of tasks to be performed by each workstation, a logical order to assemble the finished item, and the speed of each process.

Examples of processes that use assembly line layout include the automobile industry, wine-making industry, credit card processing, sandwich shops, paper manufacturers, and insurance policy processing.

There are a number of steps required in the line balancing process, as follows:

1. Identify tasks and immediate predecessors in the assembly line process.
2. Determine output rate required of the final item.
3. Determine cycle time, which is the longest time that an item can be at any one workstation. (Note that a workstation can be made up of one or more individual tasks or processes.)
4. Compute the theoretical minimum number of workstations.
5. Assign individual tasks to workstations. (That is, balance the line to make sure that the process is flowing smoothly and that no bottlenecks slow up the process.)
6. Compute efficiency and idle time and balance or eliminate any delays or bottlenecks identified.

Assembly Line Balancing Example

Step 1: Identify tasks and immediate predecessors (perfume spray filling and packaging line example in Table 16.1).

Table 16.1   Perfume and Packaging Line Balancing Example: Tasks and Predecessors

Work Element	Task Description	Immediate Predecessor	Task Time (Seconds)
A	Fill bottle	None	2
B	Place spray stem in bottle	A	3
C	Cap bottle	B	4
D	Crimp cap	C	3
E	Label bottle	D	3
F	Fold box	D	6
G	Pack bottle in box	E, F	5
H	Shrink-wrap box	G	5
		Total task time	31

Step 2: Determine output rate of final item (for example, 600 bottles per hour with 8 hours per shift).

Step 3: Determine cycle time (the amount of time each workstation is allowed to complete its tasks).

$$\text{Cycle time (sec./unit)} = \frac{\text{available time (sec./day)}}{\text{desired output per day}} = \frac{(60 \text{ min./hr.} \times 60 \text{ sec./min.}) *8\text{hrs.}}{4800 \text{ units}} = 6 \text{ sec./unit}$$

The throughput or capacity of this process is limited by the bottleneck task (the longest task in a process), which can be calculated as follows:

$$\text{Maximum output} = \frac{\text{available time}}{\text{bottleneck task time}} = \frac{360 \text{ sec./hr.}}{6 \text{ sec./unit}} = 60 \text{ units/hr.}$$

Step 4: Compute the theoretical minimum number of stations (that is, the number of stations needed to achieve 100% efficiency where every available second is used).

$$TM = \frac{\sum (\text{task times})}{\text{cycle time}} = \frac{31 \text{ seconds}}{6 \text{ sec/station}} = 5.2 \text{ or } 6 \text{ stations}$$

Note that you should always round up when calculating the number of workstations.

Step 5: Assign tasks to workstations.

Start at the first station and choose the longest eligible task following precedence relationships (that is, A must precede B, G must follow both E and F, and so on; see Table 16.2). Continue adding the longest eligible task that fits without going over the desired cycle time.

Once no additional tasks can be added within the desired cycle time, assign the next task to the following workstation until finished.

Table 16.2 Perfume and Packaging Line Balancing Example: Assign Tasks to Workstations

Workstation	Eligible Task	Selected Task	Task Time (6 Max/WS)	Idle Time
1	A	A	2	4
	B	B	3	1
2	C	C	4	2
3	D	D	3	3
	E	E	3	0
4	F	F	6	0
5	G	G	5	1
6	H	H	5	1

Step 6: Compute efficiency and balance delay.

The percent efficiency is the ratio of total task times divided by the number of workstations times the largest assigned cycle time (6 seconds for workstation 4).

$$\text{Efficiency (\%)} = \frac{\Sigma t}{NC} = \frac{31 \text{ sec.}}{6 \text{ stations} \times 6 \text{ sec.}}(100) = 86.1\%$$

Balance delay is the percentage by which the assembly line falls short of 100%.

$$\text{Balance delay} = 100\% - 86.1\% = 13.9\%$$

Work Cell Staffing and Balancing

As discussed earlier, a work cell reorganizes people and machines into groups to focus on single products or product groups that have similar characteristics and process steps or tasks. To justify a cell, there must be sufficient volume, but the cells can be reconfigured as design or volume changes.

When staffing and balancing a work cell, you must first determine the *Takt time*, which the rate at which a finished product needs to be completed to meet customer demand. It is computed as (Total work time available / Units required).

It is also important when setting up a work cell to determine staffing needs. That calculation is (Total operation time required / Takt time), which will determine the number of operators required for the cell.

Work Cell Staffing and Balancing Example

If a picture frame manufacturer requires 650 frames of a specific size per day to be produced and is running an 8-hour shift, and if the manufacturing process requires the operations (and time required) shown in Figure 16.3, we can then calculate the cell Takt time, number of workers required, as well as identify any bottlenecks in the process.

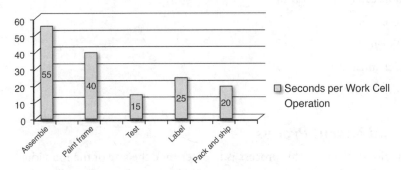

Figure 16.3 Work cell staffing and balancing example

The Takt time in this case is as follows:

(8 hrs × 60 mins) / 650 units = .74 min = **Produce 1 unit every 44 seconds**

The workers required for the cell are as follows:

155 seconds / 44 = **3.5 or 4** (round up)

By calculating the Takt time and staffing requirements of the work cell, we can now also see whether there is any imbalance in the operations caused by a bottleneck.

In this example, the assembly operation, which takes 55 seconds per picture frame (see Figure 16.3), is the bottleneck, as our Takt time is 44 seconds. The result of this is that the operations downstream will either be waiting on the assembly operation or we will have to run overtime in assembly to keep the other processes running. This will incur both higher costs and increased inventory carrying costs.

There are a variety of options to relieve this bottleneck besides running overtime, such as cross-training of operators to shift them to assembly, buying faster or additional assembly equipment, and so on.

Warehouse Design and Layout Principles

Some basic guidelines specifically apply to warehouse layout and design. In general, they include the following: use one-story facilities where possible, always try to move goods in a straight line, use the most efficient materials handling equipment, minimize aisle space, and use the full building height. (That is, land is expensive; many buildings can go as high as 65 feet.)

When designing a warehouse, other things you need to consider include the following:

- Cubic capacity utilization
- Protection
- Efficiency
- Mechanization
- Productivity

Design and Layout Process

The first major decision in this process is to determine the size of the individual warehouse facility. In addition to specifying the space needed for storage and handling, a location may be needed for processing rework and returns. All warehouses require office space for administrative and clerical activities, so space must be planned for these and other miscellaneous requirements.

The calculation to determine storage space requirements starts with a demand forecast for the facility in question.

After you have arrived at that, you then need to determine each item's order quantity (inbound and outbound), which is converted from units into cubic footage requirements to determine not only storage requirements but also other functional area requirements such as for order picking, shipping, receiving, office space, and so on, as mentioned previously (see Figure 16.4).

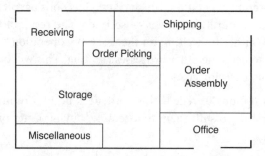

Figure 16.4 Warehouse design and layout considerations

Some of the most important functions of a warehouse actually occur at the receiving and shipping docks, so it is critical not to neglect these areas during this process.

You should consider the materials received and shipped to help determine dock bay requirements and configuration as well as staging area requirements. There are other miscellaneous requirements for the dock areas that need to be considered, such as office space, receiving hold area, trash disposal, empty pallet storage, a trucker's lounge area, and even yard space for vehicles outside.

Of course, a lot of thought needs to go into storage space planning, which requires you to do the following:

- **Define the materials to be stored:** Define what and how much material will be stored and how the materials are to be stored.

- **Select a storage philosophy:** Consider whether selecting a fixed location is necessary. If so, determine what specific location each individual SKU is stored in (even if that means that location may be empty sometimes). If a random location storage philosophy will be used instead, determine where any SKU may be assigned to any available storage location.

- **Consider space requirements for aisle space and honeycombing storage:** It is necessary within a storage area to allow accessibility to the material being stored, so an aisle space allowance (that is, a percentage) must be calculated. The amount of aisle allowance depends on the storage method, which determines the number of aisles required and the material handling method, which in turn helps to determine the size of aisles.

  Thought must also be given to an effect known as *honeycombing storage*, which is a percentage allowance of storage space lost whenever a storage location is only partially filled with material and may occur both horizontally and vertically. The unoccupied area within the storage location is honeycombing space.

In addition, there must be an allowance for growth and adequate aisle space for materials handling equipment.

Technology

As mentioned earlier in this chapter, the design and layout of an office or warehouse can be determined manually by hand, with a spreadsheet, or for more optimal results, by using packaged software such as Factory Flow, Proplanner, and CRAFT.

Careers

Whereas other functions like industrial engineering may have more say in terms of the details of warehouse design and layout, supply chain management has great input, if not the final say, for most, if not all, of the layout decisions mentioned in this chapter.

At this point, we have covered the major functional areas of supply chain and logistics management in terms of planning and managing them. Next, we look at how to control these processes.

PART VI

Supply Chain and Logistics Measurement, Control, and Improvement

17

Metrics and Measures

As the saying goes, "If you don't measure something, you can't manage (or improve) it." This was never truer than in the supply chain and logistics management field. As discussed throughout this book, an assortment of tradeoffs exist in a supply chain (for example, cost versus service), which must be counterbalanced against each other to be successful for the long term.

It is important to match your supply chain performance measures to fit your company's mission and strategy, keeping in mind that performance measures can affect the behavior of managers and employees.

It is also vital to target and measure supply chain performance to meet customer expectations, improve supply chain capability, improve asset performance, motivate the workforce, and provide stakeholders with a satisfactory return on their investment.

Although technology today makes it much easier to gather and analyze data, there is a lot more of it available, making it all the more important to measure only the right things and to avoid wasted effort; otherwise, you might fall under the dreaded phenomena of *paralysis by analysis*. The results of analysis should be used effectively.

Measurement and Control Methods

To have an efficient and effective supply chain requires a set of standards to compare to actual performance. These standards are referred to as metrics.

There are a variety of established metrics for the supply chain, such as those defined by the SCOR model mentioned in Chapter 1, "Introduction," but determining the appropriate metrics for your organization can be a complex problem.

Selecting and measuring the wrong set of metrics can lead your company to follow the wrong goals because metrics tend to drive behavior.

In supply chain and logistics management, cost metrics need to focus on the entire extended supply chain (internal and external), not on just one function or one link.

The Evolution of Metrics

Historically, businesses would focus on manufacturing costs as a measure of efficiency. That was eventually extended to transportation costs in the 1970s. In the 1980s and 1990s, this view was expanded to look at the broader performance of distribution and logistics costs, which were supplemented with more meaningful performance indicators such as the delivery rate and percentage of order fulfillment as a more customer-based focus emerged.

The advent of the global supply chain, the Internet, and enterprise resource planning (ERP) systems allowed organizations to take an even broader view of both their extended supply chains and more easily gather, measure, and analyze cost and service information.

Today, relying on traditional supply chain execution systems is becoming increasingly more difficult, with a mix of global operating systems, pricing pressures, and ever-increasing customer expectations. There are also recent economic impacts such as rising fuel costs, the global recession, supplier bases that have shrunk or moved offshore, as well as increased competition from low-cost outsourcers. All of these challenges potentially create waste in your supply chain. That's where data analytics comes in.

Data Analytics

Data analytics is the science of examining raw data to help draw conclusions about information. When applied to the supply chain, it is often described as *supply chain analytics*. It is used in many industries to allow companies and organization to drive insight, make better business decisions and actions, as well as the sciences to verify (or disprove) existing models or theories.

One way to look at data analytics is to break it into four categories:

- **Descriptive analytics:** Uses historical data to describe a business; also described as business intelligence (BI) systems. In supply chain, descriptive analytics help to better understand historical demand patterns, to understand how product flows through your supply chain, and to understand when a shipment might be late.

- **Diagnostic analytics:** Once problems occur in the supply chain, an analysis needs to be made of the source of the problem. Often this can involve analysis of the data in the systems to see why the company was missing certain components or what went wrong that caused the problem.

- **Predictive analytics:** Uses data to predict trends and patterns; often associated with statistics. In the supply chain, predictive analytics could be used to forecast future demand or to forecast the price of a product.

- **Prescriptive analytics:** Using data to select an optimal solution. In the supply chain, you might use prescriptive analytics to determine the optimal number and location of distribution centers, set your inventory levels, or schedule production.

Traditional measures tend to be based on historical data and not focused on the future, do not relate to strategic, nonfinancial performance goals such as customer service and product quality, and do not directly tie to operational effectiveness and efficiency.

Measurement Methods

A number of measurement methods have been developed or enhanced in recent times, including the following:

- **The balanced scorecard:** A strategic planning and management system that aligns business activities to the vision and strategy of the organization, improves internal and external communications, and monitors organization performance against strategic goals. It adds strategic nonfinancial performance measures to traditional financial metrics to give managers and executives a more *balanced* view of organizational performance.

- **The Supply Chain Council's SCOR model:** Metrics in this model provide a foundation for measuring performance and identifying priorities in supply chain operations.

- **Activity-based costing (ABC):** A costing and monitoring methodology that identifies activities in an organization and assigns the cost of each activity with resources to all products and services according to actual consumption by each product and service.

- **Economic value analysis (EVA):** The value created by an enterprise, basing it on operating profits in excess of capital utilized (through debt and equity financing). These types of metrics can be used to measure an enterprise's value-added contributions within a supply chain (not as useful for detailed supply chain measurements).

No matter, which method you use, it is a good idea that the metrics are consistent with overall corporate strategy, focused on customer needs, and that expectations are prioritized and focused on processes and not functions.

Furthermore, metrics should be implemented consistently throughout the supply chain, with actions and rationale communicated to everyone.

It is also a good idea to use some kind of balanced approach when selecting and developing metrics, using precise costs to measure improvement (in detail and in the aggregate), which can be greatly aided by the use of technology.

Measurement Categories

Each method has its own specifics, but it is important to at least include the measurement categories of time, quality, and cost in one form or another:

- **Time:** Includes on-time delivery and receipt, order cycle time, and variability and response time.

- **Quality:** Measures customer satisfaction, processing and fulfillment accuracy, including on-time, complete, and damage-free order delivery, as well as accurate invoicing. Also includes planning (including forecasting) and scheduling accuracy.

- **Cost:** This category includes financial measurements such as inventory turns, order-to-cash cycle time, and total delivered costs broken up by cost of goods, transportation, carrying, and material handling costs.

Another way of looking at supply chain measurement categories is in terms of where they will be applied:

- **Strategic level:** Measures include lead time against industry norm (that is, benchmarking, as discussed later in this chapter), quality level, cost-saving initiatives, and supplier pricing against market.

- **Tactical level:** Measures include the efficiency of purchase order cycle time, booking-in procedures, cash flow, quality assurance methodology, and capacity flexibility.

- **Operational level:** Measures include ability in day-to-day technical representation, adherence to developed schedule, ability to avoid complaints, and achievement of defect-free deliveries.

Now let's look at several of the major models for measuring the supply chain: the balance scorecard approach and the SCOR model.

Balanced Scorecard Approach

When using the balanced scorecard approach to the supply chain, you need to consider it from four different perspectives:

- **Customer perspective:** How do customers see us?

- **Internal business perspective:** What must we excel at?

- **Financial perspective:** How do we look to shareholders?

- **Innovation and learning perspective:** Can we continue to improve and create value?

Next we must put it in a framework that integrates the organization's overall goals and strategies with selected supply chain metrics in the categories of customer service, operations, and finance (see Figure 17.1).

It is always critical that supply chain and logistics metrics connect to the overall business strategy. For example, if a company uses a cost strategy, as described in Chapter 1, a financial metric like inventory turnover would be critical. If they employ a response strategy, inventory turns might not be as critical as, say, delivery time.

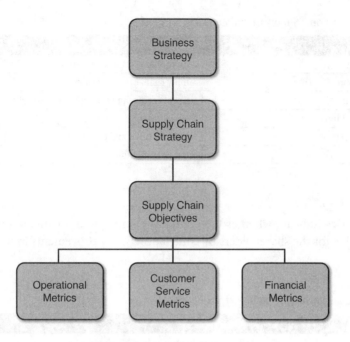

Figure 17.1 Supply chain metrics framework

Customer Service Metrics

Customer service metrics indicate a company's ability to satisfy the needs of customers by meeting customer's needs on a timely basis and creating exceptional values to the customers (see Table 17.1).

Table 17.1 Customer Service Supply Chain Metrics

Goals	Measures
Flexible response	Number of choices and average response time
Product/service innovation	Customer contact points and product finalization points
Customer satisfaction	Order-fulfillment rate
Customer value	Customer profitability
Delivery performance	Delivery speed and reliability

Operational Metrics

Operational metrics come from internal processes, decisions, and actions needed to meet or exceed customer expectations. They are drivers of future financial performance (see Table 17.2).

Table 17.2 Operational Supply Chain Metrics

Goals	Measures
Waste reduction	Supply chain cost of ownership
Time compression	Supply chain cycle efficiency
Unit cost reduction	Percentage of supply chain target cost achieved
Product/process innovation	Product finalization point
Inventory management	Inventory turns and days of inventory
Supplier performance	Supplier evaluations

Financial Metrics

Financial metrics indicate whether the company's strategy, implementation, and execution create value for the shareholders by contributing to improvements in profitability (see Table 17.3).

Table 17.3 Financial Supply Chain Metrics

Goals	Measures
Profit margins	Profit margin by supply chain partners
Cash flows	Cash-to-cash cycle on receivables and payables
Revenue growth	Customer growth and profitability
Return on assets	Return on supply chain asset
Return on equity	Return on supply chain equity

SCOR Model

Another, relatively newer approach to measuring the supply chain is the SCOR model that was discussed in Chapter 1. To recap, the SCOR model was designed to help companies to communicate, compare, and learn from competitors and companies both within and outside of their industry. It measures an organization's supply chain performance and the effectiveness of and supply chain improvements and can also help to test and plan future process improvements.

The SCOR model contains over 200 process elements, 550 metrics, and 500 best practices, including risk and environmental management, and is organized around the 5 primary management processes of plan, source, make, deliver, and return (see Figure 17.2).

Figure 17.2 The SCOR model

The model is based on three major pillars:

- **Process modeling:** To describe supply chains that are very simple or very complex using a common set of definitions, SCOR provides three levels of process detail. Each level of detail assists a company in defining scope (level 1), configuration or type of supply chain (level 2), and process element details, including performance attributes (level 3). Below level 3, companies decompose process elements and start implementing specific supply chain management practices. It is at this stage that companies define practices to achieve a competitive advantage and adapt to changing business conditions.

- **Performance measurements:** SCOR metrics are organized in a hierarchical structure. Level 1 metrics are at the most aggregated level and are typically used by top decision makers to measure the performance of the company's overall supply chain. Level 2 metrics are primary, high-level measures that may cross multiple SCOR processes.

- **Best practices:** Once the performance of the supply chain operations has been measured and performance gaps identified, it becomes important to identify what activities should be performed to close those gaps. More than 430 executable practices derived from the experience of Supply Chain Council (SCC) members are available.

Also, as mentioned in pillar 1 earlier, SCOR *levels* range from broadest to narrowest and are defined as follows:

- **Level 1. Scope:** Defines business lines, business strategy, and complete supply chains
- **Level 2. Configuration:** Defines specific planning models such as *make to order* (MTO) or *make to stock* (MTS), which are basically process strategies
- **Level 3. Activity:** Specifies tasks within the supply chain, describing what people actually do

- **Level 4. Workflow:** Includes best practices, job details, or workflow of an activity
- **Level 5. Transaction:** Specific detail transactions to perform a job step

Furthermore, all SCOR metrics have five key strategic performance attributes. A performance attribute is a group of metrics used to express a strategy. An attribute itself cannot be measured; it is used to set strategic direction.

The five strategic attributes are as follows:

- **Reliability:** The ability to deliver on-time, complete, in the right condition and packaging and with the correct documentation to the right customer
- **Responsiveness:** The speed at which products and services are provided
- **Agility:** The ability to change (the supply chain) to support changing (market) conditions
- **Cost:** The cost associated with operating the supply chain
- **Assets:** The effectiveness in managing assets in support of demand satisfaction

Table 17.4 shows some strategic metrics for level 1.

Table 17.4 SCOR Model Strategic Metrics

Performance Attribute	Sample Metric	Calculation
Supply chain reliability	Perfect order fulfillment	Total perfect orders / Total number of orders
Supply chain responsiveness	Average order-fulfillment cycle time	Sum of actual cycle times for all orders delivered / Total number of orders delivered
Supply chain agility	Upside supply chain flexibility	Time required to achieve an unplanned X% increase in delivered quantities
Supply chain costs	Supply chain management costs	Cost to plan + Cost to source + Cost to deliver + Cost to return
Supply chain asset management	Cash-to-cash cycle time	Inventory days of supply + Days of receivables outstanding – Days of payables outstanding

At lower levels of detail, the SCOR model sets a variety of specific, standard measures that can be used by an organization. Table 17.5 lists some examples under the performance attribute of *supply chain responsiveness* for the *deliver* component of the SCOR model.

Table 17.5 Supply Chain Responsiveness Delivery Cycle Time Example

	Supply Chain Responsiveness
RS.2.3	**Delivery cycle time**
RS.3.16	Build loads cycle time
RS.3.18	Consolidate orders cycle time
RS.3.46	Install product cycle time
RS.3.51	Load product and generate shipping documentation cycle time
RS.3.95	Pack product cycle time
RS.3.102	Receive and verify products by customer cycle time

Supply Chain Dashboard and KPIs

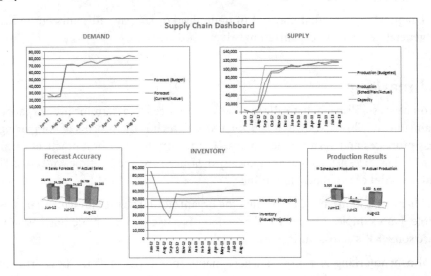

Source: S&OP Excel Template, published with permission of Logistics Planning Associates, LLC

One way to measure, analyze, and manage supply chain performance is with the use of a *dashboard*. The dashboard can range from data that is manually collected and put into a spreadsheet with some graphs, to a more automated visual dashboard generated by an ERP system.

A supply chain dashboard helps in decision making by visually displaying in real time (or close to it) leading and lagging indicators in a supply chain process perspective. It can help to you visualize trends, track performance targets, and understand the most critical issues facing your company's supply chain.

Indicators

Metrics used in performance dashboards are typically called *key performance indicators* (KPIs). Having a standardized set of KPIs allows you to review supply chain operations efficiently across regions, business units, and plants (and even brands and channels).

KPIs usually fall into one of three categories:

- **Leading indicators:** Have a significant impact on future performance by measuring either current state activities (for example, the number of items produced today) or future activities (for example, the number of items scheduled for production this week)

- **Lagging indicators:** Measures of past performance, such as various financial measurements, or in the case of the supply chain, measurements in areas such as cost, quality, and delivery

- **Diagnostic:** Areas that may not fit under lead or lagging indicators but indicate the general health of an organization (Myerson, 2012)

Benchmarking

After you have established what KPIs to measure, you need to determine how to gauge yourself against them. This is known as *benchmarking*, which is the process of comparing one's business processes and performance metrics to industry bests or best practices from other companies. The dimensions that are typically measured are quality, time, and cost. There is both internal and external benchmarking.

Internal benchmarking can be used when the organization is large enough and data is fairly accessible. For example, a company like General Electric with over 100 business units can compare some KPIs across businesses.

External benchmarking is a process where management identifies the best firms in their industry, or in another industry where similar processes exist, comparing the results and processes of those *target* companies to their own results and processes. This allows them to learn how well the targets perform and, more importantly, the *best practice* business processes that help to explain why these firms are successful.

The process of selecting best practices to use as a standard for performance involves the following general steps:

1. **Determine what to benchmark.** What processes are most important to measuring the success of your supply chain? Can be based on your company's overall and supply chain strategies, goals, and objectives. In some cases, you may want to be best in class, and in others, being average may be just fine.

2. **Form a benchmark team.** Select those with some skin in the game.

3. **Identify benchmarking partners.** Should be a combination of those involved in the day-to-day processes as well as those affected and management (can also include customers and suppliers).

4. **Collect and analyze benchmarking information.** There are many sources of best practice metrics. In many cases, companies can pay for this information from consultants or through associations that they belong to.

5. **Take action to match or exceed the benchmark.** This involves process improvement, the topic of the next chapter.

18

Lean and Agile Supply Chain and Logistics

I t is the goal of every business to be both profitable and efficient. Over the past 25 years or so, one method has emerged as a way to focus on customer needs to improve processes and profitability; it is known as Lean, which is discussed in this chapter, and can be applied throughout the supply chain and logistics function.

Lean and Waste

By definition, Lean is a team-based form of continuous improvement to identify and eliminate waste through a focus on exactly what the customer wants. Lean supplies the customer with exactly what they want when they want it, through continuous improvement. Lean is driven by the *pull* of the customer's order.

This is in contrast to just-in-time (JIT), which is a philosophy of continuous and forced problem solving that supports Lean production. JIT is one of the many tools of Lean. When implemented as a comprehensive strategy, JIT and Lean sustain competitive advantage and result in greater returns for an organization.

To successfully execute a Lean strategy, the supply chain needs to also be very agile, meaning that it must be designed to be flexible and fast when dealing with *unique* products with unpredictable demand.

So, not surprisingly, that's where the idea of having a *hybrid* strategy comes into play. A hybrid supply chain strategy is a combination of Lean and agile concepts, where a manufacturer operates with flexible capacity that can meet surges in demand along with a postponement strategy, where products are partially assembled to a forecast and then completed to the actual order when and even where it arrives.

In the end, your supply chain strategy (and capabilities) must support your organization's overall strategy, and typically, in today's world, that means being Lean and agile.

This chapter describes the methodology and some of the tools used to enable a Lean and agile supply chain and logistics function.

History of Lean

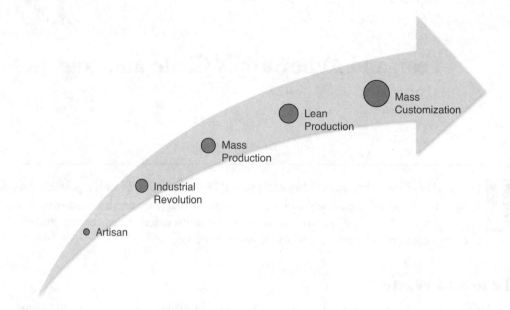

Figure 18.1 History of Lean

Early concepts like *labor specialization* (Smith), where an individual was responsible for a single repeatable activity, and *standardized parts* (Whitney) helped to improve efficiency and quality.

At the turn of the 20th century, the era of scientific management arrived, where concepts such as *time and motion studies* (Taylor) and *Gantt charts* (Gantt) allowed management to measure, analyze, and manage activities much more precisely.

During the early 1900s, the era of mass production had arrived. Concepts such as the assembly line, economies of scale (producing large quantities of the same item to spread fixed costs), and statistical sampling were utilized. Today, this is referred to as a *push process*, which is the opposite of *demand pull* (by the customer) used in a Lean philosophy.

Lean, originally applied to the manufacturing industry, was developed by the Japanese automotive industry, largely Toyota, while rebuilding the Japanese economy after World War II. Material was scarce, and they realized that to compete with the U.S. auto companies, they would have to work smarter.

The concept of Lean was little known outside Japan until the 1970s (generally known as *JIT* as the actual term *Lean* didn't come about until the 1990s). England had early experience with Lean manufacturing from the Japanese automotive plants in the United Kingdom.

Up until the 1990s, only the automotive industry had adopted Lean manufacturing, and that was primarily on the shop floor. Since then, it has spread into aerospace and general manufacturing, consumer electronics, healthcare, construction, and, more recently, to food manufacturing and meat processing, in addition to other processes such as the administrative and support functions as well as the supply chain.

In today's global economy, companies source product and material worldwide, looking for the best quality at the lowest cost. E-commerce and enterprise resource planning (ERP) systems have made for easy entry to the global economy for smaller companies, as well, allowing them to compete against much larger competitors.

This has led to the concept of *mass customization*, which is the ability to combine low per-unit costs of mass production with the flexibility associated with individual customization (for example, Dell computers, which can configure, assemble, test, and ship your customized order within 24 hours).

Value-Added Versus Non-Value-Added Activities

To understand the Lean concept of *waste*, it is first important to understand the meaning of value-added versus non-value-added activities.

Any process entails a set of activities. The activities in total are known as *cycle* or *lead time*. Lead time required for a product to move through a process from start to finish includes queues/waiting time and processing time.

The individual activities or work elements that actually transform inputs (for example, raw materials) to outputs (for example, finished goods) are known as *processing time*. In general, processing adds value from the customer's standpoint. Processing time is the time that it takes an employee to go through all of their work elements before repeating them. It is measured from the beginning of a process step to the end of that process step.

If we think of a simple example such as taking raw lumber and making it into a pallet of 2x4s, the value added to the customer is the actual processing that transforms the raw lumber into the final pallet of 2x4s. This would include activities such as washing, trimming, cutting, and so on, and are a relatively small part of the cycle time. (That is, it might only take 1 hour to process the raw material into a finished pallet, but the entire cycle time may be 1 week.)

In Lean terms, the non-value-added time is actually much greater than just the lead time. We include current inventory *on the floor* (that is, raw, work in process [WIP] and finished goods), and using a calculated Takt time for a specific *value stream* (a single or family of products or service, as discussed later in this chapter), convert those quantities to days of supply. Doing so can expand the non-value-added time from days to weeks (or even months).

It is common for many processes (or value streams) to only have 5% to 10% value-added activities. However, there are some non-value-added but necessary activities, such as regulatory, customer required, and legal requirements, which although they do not add value, are considered waste. Because they are necessary, we cannot eliminate them, but should try to apply them as efficiently as is possible.

It is fairly normal for management to focus primarily on speeding up processes (often value-added ones such as the stamping speed on a press). From a Lean perspective, the focus moves to non-value-added activities, which in some cases may even result in slowing down the entire process to balance it, and removes bottlenecks and increases flow.

Waste

In Lean terms, non-value-added activities are referred to as *waste*. Typically, when a product or information is being stored, inspected or delayed, waiting in line, or is defective, it is not adding value and is 100% waste.

These wastes can be found in any process, whether it is manufacturing, administrative, supply chain and logistics, or elsewhere in your organization.

Listed here are the eight wastes. One easy way to remember them is that they spell TIM WOODS:

- **Transportation:** Excessive movement of people, products, and information.
- **Inventory:** Storing material or documentation ahead of requirements. Excess inventory often covers for variations in processes as a result of high scrap or rework levels, long setup times, late deliveries, process downtime, and quality problems.
- **Motion:** Unnecessary bending, turning, reaching, and lifting.
- **Waiting:** For parts, information, instructions, and equipment.
- **Overproduction:** Making more than is immediately required.
- **Overprocessing:** Tighter tolerances or higher-grade materials than are necessary.
- **Defects:** Rework, scrap, and incorrect documentation (that is, errors).
- **Skills:** Underutilizing capabilities of employees, delegating tasks with inadequate training.

There are a variety of places to look for waste in your supply chain and logistics function.

One way to consider where waste might exist in the supply chain is in terms of the SCOR model:

- **Plan:** It all starts (and ends) with a solid sales & operations planning (S&OP) process (or lack thereof). If there is not one in an organization, there is probably plenty of waste in the supply chain.
- **Source:** Because purchasing accounts for approximately 50% (or more) of the total expenditures, using sound procurement approaches discussed earlier in this book

such as economic order quantity (EOQ) and the use of just-in-time (JIT) principles (including vendor-managed inventory, or VMI) can go a long way toward reducing waste, especially excess inventory. Partnerships, collaborations, and joint reviews with suppliers can also help to identify and reduce waste.

- **Make (and store):** Activities such as (light) manufacturing, assembly, and kitting, much of which is done in the warehouse or by a third-party logistics supplier (3PL) these days, can have a huge impact on material and information flow, impacting productivity and profitability.

Within the warehouse, waste can be found throughout the receiving, putaway, storage, picking, staging, and shipping processes, including the following:

- Defective products, which create returns
- Overproduction or over shipment of products
- Excess inventories, which require additional space and reduce warehousing efficiency
- Excess motion and handling
- Inefficiencies and unnecessary processing steps
- Transportation steps and distances
- Waiting for parts, materials, and information
- Information processes

Each step in the warehousing process should be examined critically to see where unnecessary, repetitive, and non-value-added activities might be so that they may be eliminated:

- **Deliver:** Transportation optimization (especially important with high fuel prices). This includes routing, scheduling, and maintenance, among other things.
- **Return:** Shipping mistakes, returns, and product quality and warranty issues often ignored or an afterthought.

There are a variety of Lean tools available, as discussed briefly in this chapter. However, it is most important to first have a Lean culture and support that is conducive to success, because Lean is more of a journey than any individual project.

Lean Culture and Teamwork

To be successful for the long term, any type of program has some key success factors (KSF). In the case of Lean, they include the following:

- Train the entire organization and make sure that everyone understands the Lean philosophy and understands that it may be a cultural change for the organization.

- Ensure that top management actively drive and support the change with strong leadership.

- Everyone in the organization should commit to make it work.

- Find a good, experienced change agent as the *champion.*

- Set a kaizen (that is, continuous improvement project) agenda and communicate it and involve operators through empowered teams.

- Map value streams.

- Apply Lean tools and begin as soon as possible with an important and visible activity.

- Integrate the supporting functions and build internal customer and supplier relationships.

Lean Teams

Teamwork is essential for competing in today's global arena, where individual perfection is not as desirable as a high level of collective performance.

In knowledge-based enterprises, teams are the norm rather than the exception. A critical feature of these teams is that they have a significant degree of empowerment, or decision-making authority.

There are many different kinds of teams: top management teams, focused task forces, self-directed teams, concurrent engineering teams, product/service development and/or launch teams, quality improvement teams, and so on.

It is no different in Lean. As a result, it is important to establish Lean teams that can develop a systematic process that consistently defines and solves problems utilizing Lean tools.

Lean teams are a great way to share ideas and create a support system helping to ensure better *buy-in* for implementation of improvements.

Successful teams realize the *power of teamwork* and teamwork culture and that the goal is more important than anyone's individual role. However, teams must be in a *risk-free* environment but have leadership, discipline, trust, and the tools and training to make things happen.

To make teamwork happen, it is important that executive leaders communicate the clear expectation that teamwork and collaboration are expected and that the organization members talk about and identify the value of a teamwork culture. Teamwork should be rewarded and recognized, and important stories and folklore that people discuss within the company should emphasize teamwork.

Kaizen and Teams

The work *kaizen*, which literally means *improvement* in Japanese, refers to activities that continually improve all functions and involve all employees.

Kaizen events are a big part of a successful Lean program. A typical kaizen event involves a team of people for a period of 3 to 10 days. They typically focus on a working (or proposed) process with the goal of a rapid, dramatic performance improvement. Typically, the event starts with training on topic of the event to ensure common understanding.

Team and Kaizen Objectives

Once teams have been established and basic training has begun to start a Lean philosophy, it is important for any projects undertaken by teams to have proper objectives to ensure success. So, you need to ask questions such as the following:

- What is the customer telling you in terms of the cost, service, and quality of your products/services?
- What objectives and goals have been established by your company to address market needs?
- What processes immediately impact the performance of these products and services?
- Who needs to support this effort?
- How can the business objectives be used to garner support?

By asking these types of questions, it is not hard to link Lean projects to overall and functional objectives and metrics for improvement.

Value Stream Mapping

Before discussion specific Lean tools, it is first important to understand the current processes to be examined. Lean has a visual, relatively high, and broad-level method for analyzing a current state and designing a future state for the series of activities (both value-added and non-value-added) required to produce a single or family of products or services for a customer known as a *value stream map* (VSM).

A VSM is typically labeled a *paper and pencil tool*, although it may be constructed digitally and is a value management tool designed to create two separate visual representations (that is, maps).

The first map illustrates how data and resources move through the value stream during the production process, and is used to identify wastes, defects, and failures (see Figure 18.2); the second map, using data contained in the first, illustrates a future state map of the same value stream with any waste, defects, and failures eliminated (see Figure 18.3).

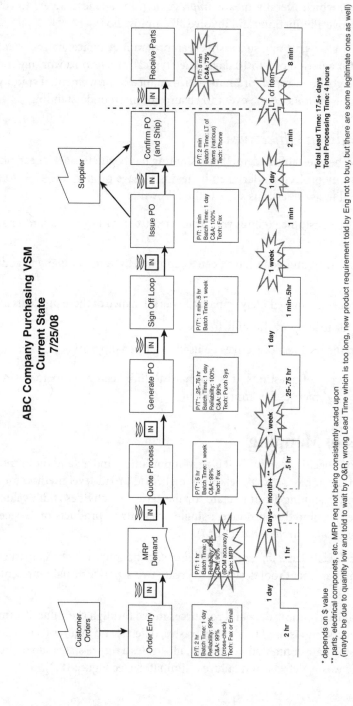

Figure 18.2 Current state value stream map

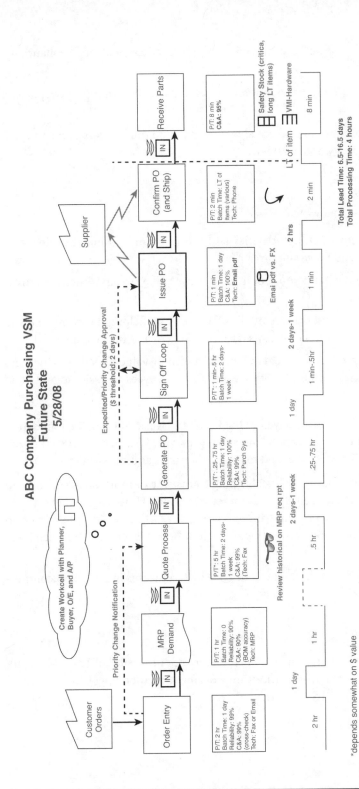

Figure 18.3 Future state value stream map

The two maps are used to create detailed strategic and implementation plans to enhance the value stream's performance.

A VSM is usually one of the first steps your company should take in creating an overall lean initiative plan.

Developing a visual map of the value stream allows everyone to fully understand and agree on how value is produced and where waste occurs.

VSM Benefits

Benefits of VSM include the following:

- Highlights connections among activities and information and material flow that impact the lead time of your value stream.

- Helps employees understand your company's entire value stream rather than just a single function of it.

- Improves decision-making process of all work teams by helping team members to understand and accept your company's current practices and future plans.

- Allows you to separate value-added activities from non-value-added activities and then measure their lead time. Provides a way for employees to easily identify and eliminate areas of waste.

As was previously discussed and is displayed in Figure 18.4's House of Lean, the foundation for a Lean enterprise (including the supply chain and logistics areas) is to have a Lean culture and infrastructure as well as a way to set and measure objectives and performance.

After performing basic Lean training and establishing teams, one or more value stream map studies are typically performed to identify areas for improvement. However, it is not uncommon for teams to brainstorm to come up with ideas for future kaizen events in addition to or instead of value stream mapping events. (Figure 18.5 shows an example of a kaizen form for brainstorming ideas.)

Next, we will look at some of the tools that can be used by Lean teams in pursuit of continuous improvement.

Lean Tools

There are a variety of tools that are now included under the "umbrella" of Lean. Some are unique to Lean and others come from various improvement methodologies such as Total Quality Management (TQM) and Six Sigma.

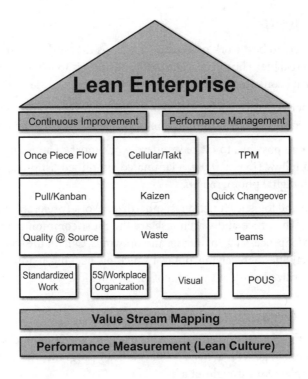

Figure 18.4 House of Lean

Cost Reduction Kaizen Implementation

Department:_____ Process for Kaizen: _____   Kaizen #:_____

Cost Center:_____   Date: _____

Approvals: Lean Champion:_____   Maint:_____   Controller:_____   GM:_____

1) Current Situation	3) Solution Activity

2) Analysis	4) Cost Reduction (Total Savings:)
	Current Proposed

Figure 18.5 Cost reduction kaizen implementation form

Standardized Work

Standardization refers to best work practices—that is, as the work is actually routinely (and best) performed in real life. The purpose of standardization is to make operations repeatable and reliable, ensuring consistently high productivity and reduced variability of output.

It ensures that all activities are safely carried out, with all tasks organized in the best known sequence using the most effective combination of people, material, machines, and methods.

It is important, where possible, to make standard work more of a *visual job aid* that is easy to understand and follow (for example, a laminated simplified list of standard instructions supplemented with digital photographs).

In the supply chain and logistics function, standardized work (preferably visual) can applied nearly everywhere. The office and warehouse are the most common places they are found and can include order processing, invoicing, and drawings. Out on the warehouse floor itself, most of the basic activities of receiving, putaway, picking, packing, loading, and shipping can benefit from standardized work in the form of visual job aids.

5S-Workplace Organization System

5S is a philosophy that focuses on effective workplace organization and standardized work procedures. It is a great, general activity to start a Lean program with because it is easy to understand and implement throughout a business.

5S simplifies your work environment and reduces waste and non-value activity while improving quality efficiency and safety. It ensures that the workplace that is clean, organized, orderly, safe, efficient, and pleasant; and it results in the following:

- Fewer accidents
- Improved efficiency
- Reduced searching time
- Reduced contamination
- Visual workplace control
- A foundation for all other improvement activities

A 5S project begins with the selection of a specific area (usually one that is fairly disorganized) and a multifunctional team that includes at least one member from the selected area.

Next, the team goes to the selected area to perform a *workplace scan*, which involves activities such as taking *before* pictures, drawing a *spaghetti diagram* showing locations of materials and equipment as well as product flow, and the performance of some kind of 5S audit. (Various forms are readily available on the Internet.)

The steps in 5S (and what the actual *S*'s stand for) are as follows:

- **Sort:** Unneeded items are identified and removed. Only needed parts, tools, and instructions remain.
- **Set in order:** Everything has a place; everything is in its place. Create visual controls to know where items belong and when they are missing as well as how much to keep on hand in the area.
- **Shine:** Do an initial spring cleaning. This can include scouring as well as some painting.
- **Standardize:** Routine cleaning becomes a way of life. Preventive maintenance is routinely performed. Standards are created to maintain the first three S's.
- **Sustain:** This is perhaps the hardest part of 5S, where it has to become a routine way of life. Root causes are routinely identified and dealt with.

Visual Controls

Simple visual signals give the operator the information to make the right decision. They are efficient, self-regulating, and worker managed.

Examples include visual job aids mentioned previously, kanbans, *andon* lights (that is, green = process working; red = process stopped), and color-coded dies, tools, pallets, and lines on the floor to delineate storage areas, walkways, work areas, and so on.

Facility Layout

Considering optimal facility layout, like standardized work, is nothing new. However, as a tool of Lean, it is focused primarily on maximizing flow and eliminating wastes such as transportation and motion. If used properly, it can result in the following:

- Higher utilization of space, equipment, and people
- Improved flow of information, materials, or people
- Improved employee morale
- Improved customer/client interface
- Increased flexibility

Batch Size Reduction and Quick Changeover

The concepts of batch size reduction and quick changeover (sometimes also referred to as *setup reduction*) are highly intertwined.

When material is *pushed* through a supply chain and operations process, you produce, store, and ship in large quantities to spread your fixed costs among a large number of items, thus minimizing your costs per unit. In *pull*, you schedule what the downstream customer actually wants, using a JIT approach. The goal is one-piece flow (or at least a reduction in batch or lot size).

Long changeovers tend to create larger batch sizes, resulting in higher inventory costs, longer lead times, and potentially larger quality issues; and that is why we focus on reducing changeover times through *setup reduction* kaizen events.

In supply chain and logistics processes, we often see the results of batching in production to cover manufacturing wastes that result in excess inventory, and in purchasing to obtain economies of scale. In addition, there is a large amount of batching of paperwork in the office, which if reduced, can encourage improved flow and getting orders out faster, resulting in a shorter order-to-cash cycle.

In warehouse operations, there are setups everywhere, including receiving, picking, staging, loading, and shipping (especially during shift startups).

Quality at the Source

Also known as *source control*, the idea with this concept is that the next step in the process is your customer, and as a result, you need to ensure perfect product for your customer.

One major technique used in source control is known as *poka-yoke*, which is the concept of using foolproof devices or techniques designed to pass only acceptable product. Poka-yokes can range from simple tools such as a *cutout* to ensure proper dimensions, to a scale at a packing station that checks the weight of an item (and if it is outside of the proper range, software would prevent a label from printing).

Quality at the source can eliminate or reduce final inspections, reduce passed-on defects, eliminate non-value-added processing, increase throughput, and increase employee satisfaction.

Quality at the source helps to reduce the total cost of quality, which looks at the true impact of defective work as it moves toward the customer. This includes the following:

- **Prevention costs:** Reducing the potential for defects (for example, poka-yokes)
- **Appraisal costs:** Evaluating products, parts, and services (for example, quality control sampling)
- **Internal failure:** Producing defective parts or service before delivery (for example, final inspection)
- **External costs:** Defects discovered after delivery (for example, returns)

Obviously, the further along a quality issue gets, the greater impact (cost and otherwise) it has on an organization.

Other techniques such as standardized work, visual workplace, and 5S are all tools of implementing quality at the source.

Point-of-Use Storage

Point-of-use storage is the storing of raw materials and supplies needed by a work area that will use them nearby. It works best if the supplier can deliver frequent, on-time, small deliveries. It can simplify the physical inventory tracking, storage, and handling processes.

Total Productive Maintenance

Total productive maintenance (TPM) is often used interchangeably with the concept of preventive maintenance. Although preventive maintenance may be involved, TPM is actually a team-based systematic approach to the elimination of equipment-related waste. It involves the charting and analysis of equipment performance to identify the root cause of problems, then implementing permanent corrective actions.

TPM is a shared responsibility that involves not only mechanics but also operators, engineers, and employees from other functional area.

Ultimately, in addition to creating *countermeasures* using techniques such as poka-yokes, TPM develops preventive maintenance plans that utilize the best practices of operators, maintenance departments, and depot service. It also involves the training of workers to operate and maintain their own machines, often referred to as *autonomous maintenance.*

Although the supply chain and logistics function might not have as many or as complicated equipment as in manufacturing, there is plenty of equipment that must run at peak performance, including forklifts and carousels in a warehouse, trucks on the road, and office equipment such as computer hardware and software.

Pull/Kanban and Work Cells

As mentioned before, a *push* system produces product, using forecasts or schedules, without regard for what is needed by the next operation, whereas a *pull* system is a method of controlling the flow of resources by indirectly linking dissimilar functions, through the use of visual controls (for example, kanbans), replacing only what has been consumed at the demand rate of the customer.

A pull system is a flexible and simple method of controlling and balancing the flow of resources and eliminates the waste of handling, storage, expediting, obsolescence, rework, facilities, equipment, and excess paperwork. It consists of processing based on actual consumption, low and well-planned work in process (paperwork), and management by sight, with improved communication.

One of the main tools in a pull system is a *kanban*, in which a user removes a standard-sized container, and a signal is seen by the producing/supplying department as authorization to replenish. The signal can be a card or even something as simple as a line on a wall.

Another Lean tool is known as a *work cell*, covered in Chapter 16, "Facility Layout Decision," which reorganizes people and machines that typically would be dispersed in various departments into a group so that they can focus on making single product or group of related items. Work cells are usually *U* shaped versus a traditional linear assembly line type of format.

Work cells require the identification of families of products or services and require a high level of training and flexibility on the part of employees and in many cases utilize poka-yokes at each station in the cell.

Work cells can be found in a variety of industries on the shop floor, in a warehouse, and in the office. In the warehouse, there may be more limited opportunities than elsewhere, but they are usually found in areas such as packaging or in value-added activities performed by 3PLs such as packaging of kits for a customer or a staging location to organize outgoing shipments.

Lean and Six Sigma

In recent years, Lean has often been combined with Six Sigma to become *Lean Six Sigma* in many companies. The concept of Six Sigma was originated by Motorola in the early 1980s, and is now used in many industries. Six Sigma attempts to improve the quality of process outputs by identifying and removing the causes of defects (errors) and minimizing variability in manufacturing and business processes, thus the term *Six Sigma*, which refers to a process that has 99.99966% of products produced free of defects.

Lean and Six Sigma are complementary because Lean uses relatively simple concepts to make improvements and covers the *entire* process or value stream, beginning with the customer end upstream to suppliers, and Six Sigma is a tool (heavily statistical) that looks at individual steps in the process and attempts to identify and remove defects and variability. In general, Lean tries to reduce waste in the production process, and Six Sigma tries to add value to the production process (Myerson, 2012).

The concept of Lean in the supply chain has been gaining increasing popularity during the past 5+ years as it is an ideal way to create an improved supply chain which is focused on adding value to the customer.

19

Outlook for Supply Chain and Logistics Management

I t is truly an exciting time to work in supply chain and logistics management in today's global economy. Over the past 25 years, the field of supply chain has risen in importance in a variety of organizations to the point where many now have "C" level supply chain executives.

Supply Chain and Logistics Career Outlook

Supply chain managers have a great impact on the success of an organization. As described in this book, they are involved in every aspect of a business, including planning, purchasing, production, transportation, storage and distribution, customer service, and beyond. Their performance helps organizations to control expenses and increase sales and profits.

Many organizations today rely on new technologies and the coordination of processes to expedite the distribution of goods. The use of computers to analyze work routines to optimize the use of available labor has led to increases in productivity. As a result, there is now a new set of integrated operations management functions, which require tech-savvy highly trained managers of supply chains, resource managers of material or manufacturing resources planning (MRP), and process and inventory control managers.

Business school graduates who are competent and well prepared, and who have solid knowledge in supply chain and logistics management, are in high demand across all industries. Supply chain management students are prepared for positions such as procurement/sourcing manager, logistics planner, supply management analyst, acquisition project analyst, marketing analyst, and sales/distribution managers. Industries such as pharmaceutical and healthcare companies are investing heavily in creating and supporting supply chains that achieve new heights of efficiency and productivity.

In fact, according to the Bureau of Labor Statistics (BLS), supply chain management is projected to be one of the fastest growing industries in the coming years, with employment increasing 83% since 2008. Furthermore, the BLS states that the employment of "logisticians" is projected to grow 22% from 2012 to 2022, much faster than the average for all occupations;

employment growth will be driven by the important role logistics plays in the transportation of goods in a global economy (source: www.bls.gov).

The Institute for Supply Chain Management's (ISM) 2011 survey showed that the average salary for supply chain management professionals is $103,664, up from $98,200 a year earlier. The average entry-level professional supply management salary is about $49,500, but the average salary of those with 5 or fewer years of experience is $83,689, up from $72,908 in 2010, an increase of nearly 15% (source: www.ism.ws).

While we know the outlook for careers in supply chain and logistics management is promising, it is also important to consider future trends in the field because they will significantly impact both education and business practices.

Trends in Supply Chain and Logistics Management

In a rapidly changing world, it is important to observe and understand trends in supply chain and logistics management, processes, and technology.

Supply Chain Trends

According to an SCM World report (scmworld.com, 2014), supply chain strategists will most likely need to change from their traditional cost reduction, standardized process views because efficiency no longer wins on its own. They will now also need to be prepared for increased uncertainty and volatility with an efficient *and* agile supply chain.

Trends that the report noted include the following:

- **Customer centricity:** Customer-centric supply chains are created by integrating point-of-sale, channel, and social data to help determine both short-term and long-term demand. Integrating this data with a view of supply enables demand management to shift customers to the most profitable solutions while maximizing customer satisfaction.

- **Digital demand and omni-channel:** Almost universally, digital demand is a major supply chain disruptor, and for some businesses, progress has been made in this area. Studies indicate that the majority of supply chain organizations continue to support direct-to-customer delivery in response to e-commerce and mobile demand, either via third parties or in-house fulfillment.

- **Distribution networks:** Customer demand volatility and the trend toward mass customization is a major factor impacting distribution and logistics network design decisions.

- **Manufacturing strategies:** Since the 1980s, outsourced manufacturing was a major focus for supply chain strategists. There is somewhat of a trend today toward a return to in-house production. This, of course, has a major impact on supply chain cost and efficient operations.

- **Design for profitability:** Many businesses still aren't totally open and collaborative in terms of new product development and launch. Because much of a product's costs are locked in during the initial design phase, there needs to be a shift in this thinking toward greater involvement of supply chain personnel, suppliers, and partners (and customers in some cases).

- **Sourcing and supplier management:** The increased importance of supplier management and partnering, as opposed to traditional transactional-focused purchasing, is similar to the increased complexity we see in manufacturing and product design.

- **Sales and operations planning:** Most companies have understand the necessity and importance of S&OP, but are light on the actual execution and long-term vision of the process.

- **Risk management:** Supply chain risk management is still a rather immature discipline, but with major weather, energy, and terrorist disruptions during the past few years, many companies have started to take it more seriously and invest in developing internal risk management capabilities.

- **Sustainability:** There has been a steady increase in the use of cost savings as a justification for investments in sustainability initiatives.

- **Transformation and talent:** A company's strategy to talent development, acquisition, and retention has to be integrated into the broader strategic plan of the supply chain organization.

Logistics Trends

Trends that may impact the future of global logistics, as indicated in an article from SupplyChainDigital.com (Nabben, 2014), include the following:

- **Growth patterns:** Growth in the logistics industry is no longer driven primarily by imports. It will come from elsewhere, and will be more fragmented, more unpredictable, and more volatile.

- **Flexibility:** Meeting consumer requirements at multiple locations with multiple transport modes at different times requires a flexible supply chain that can adapt easily to unexpected changes and circumstances.

- **Globalization:** International, mature, and emerging markets have become a part of the overall business growth strategy for many companies.

- **Near shoring:** As labor costs in Asia and transportation costs rise, increasing amounts of manufacturing are being brought closer to the end user.

- **Multichannel sourcing:** End consumers increasingly source via multiple channels, ranging from brick-and-mortar shops to e-commerce supported by the logistics industry.

- **Information technology:** The growing complexity and dynamism of supply chains requires increasingly advanced information technology solutions.

- **Continuity:** For speed to market and to reduce risk of delays, alternative transport modes and routes are required to support the continuing trend of outsourcing of logistics services.

- **Sustainability:** Customers increasingly prefer products that minimize businesses' social, economic, and environmental impact on society and enhancing positive effects.

- **Partnerships:** Manufacturers continuously search for supply chain innovations and gains through partnerships with logistic service providers.

- **End-to-end visibility:** Complete visibility of the entire supply chain aspires to achieve true demand-driven planning, allowing efficient response to changes in sourcing, supply, capacity, and demand.

- **Complexity:** Supply chains are becoming increasingly complex and dynamic, with sourcing locations being changed increasingly quickly and purchase orders becoming smaller and more frequent.

Supply Chain Leadership Trends

Companies that have strong supply chain visibility and analytics capabilities are almost twice as likely to be in the top 20% of business growth than firms that don't, according to CSC's ninth annual Global Survey of Supply Chain Progress (CSC, 2014).

Furthermore, the study identified the following "best practice" trends found in industry leaders:

- In most companies, supply chains have gradually shifted from defense and survival since the great recession of 2007 toward contributing to profitable growth.

- Supply chain leaders are more likely to put logistics and supply chain management into the hands of top officers. (that is, they are more likely to make the supply chain chief one of the C-level executives), and are twice as likely to have the supply chain manager report directly to a corporate officer.

- Supply chain visibility is a key advantage for leaders, who rated their visibility higher than the rest in the survey. This meant that they had the best visibility into their customers' sales information, promotional plans, and demand forecasts; and their suppliers' inventory, order lead times, and delivery dates.

- Leaders were found to be twice as likely as laggards to be experienced users of data analytics software, and as a result, they are able to extract more useful information from their supply chain visibility data.

- Supply chain leaders are also flexible enough to quickly change their supply chain solutions to respond to new threats or opportunities.

It was also found that businesses with high supply chain visibility had greater cost reduction.

Additional standout trends for supply chain leaders according to "The Gartner Supply Chain Top 25 for 2014" (Aronow, Hofman, Burkett, Nilles, & Romano, 2014) are as follows:

- **Understanding and supporting the fully contextualized customer:** "An enduring trait of leading companies is that customer needs and behaviors serve as the starting point for go-to-market and operational support strategies. The best of them present simple, elegant solutions to their customers, driven by conscious supply chain orchestration behind the curtain. Their center-led cultures enable consistently high-quality customer experiences tailored, where important, to local tastes."

- **A convergence of digital and physical supply chains delivering total customer solutions:** "Leading companies have moved past selling only discrete products or services to their customers and are now focused on delivering solutions. Regardless of industry, these companies want their customers to be loyal subscribers to their solutions. Several of the leading consumer product companies on this year's list are offering e-commerce subscriptions for their products, in partnership with retailers, to create a seamless multichannel experience. This approach offers convenience and privacy to end customers that would normally buy these products in a physical store and might switch to another consumer brand during any given store shopping visit."

- **Supply chain as trusted and integrated partner:** "Growth is a top priority for the C-suite in 2014, with 63 percent of senior executives picking growth as a top imperative in Gartner's 2014 CEO Survey. Leading supply chains are enabling this growth both organically and through successful M&A integration. At the same time, supply chain leaders are emerging as trusted and integrated partners to business groups. Their focus on profitable growth often leads to smarter, more conscious decision making, saving business groups from spiraling out of control in the drive to maximize revenue."

Supply Chain Technology Trends

Over the past 30 years, technology has been an enabler of supply chains, especially as the supply chain has extended globally. It is pervasive in all aspects of planning and execution, as can be seen throughout this book.

The Capgemini Consulting Institute conducted a Supply Chain Management Trend Study that looked at the impact of the latest technology trends toward SCM. They identified six major elements that they believe will have a great impact on SCM in the next few years (Schneider-Maul, 2014):

- **Emerging technologies:** The advance in this area is machine-to-machine communication supported by sensors and content information. Production processes can be accelerated with more transparency available for more informed management and operation decision making. This will lead to cost improvements based on avoidance of wasteful troubleshooting activities due to the unknown status of material.

- **Analytics and simulation:** These tools will be a major element of future supply chain control. In the future, decisions will be based more on real-time information instead of assumptions and only historical data and will therefore enable decision makers at all levels to envision future scenarios easily and quickly.

- **Supply chain segmentation:** Instead of the common "one size fits all" concepts of the past, supply chain segmentation allows companies to address customer-specific requirements. To implement a segmentation-oriented planning process, it is necessary to use simulation technology and create scenarios to define the appropriate supply chain by segment.

- **Service orientation:** This is strongly related to the segmentation trend, but also linked to the implementation of supply chain service and control centers to ensure that service level targets are met. The key to success is the centralized integration of global supply chain event information. This enables the supply chain control group or teams in the supply chain service centers to react to supply chain events quickly to make the right decisions.

- **Optimization of supply chains:** This strategic task is already supported by supply chain network optimization tools that are already available in the market. They apply scenario modeling and simulation to define the best possible supply chain configuration. In addition, procurement optimization software supports material cost and spend analysis and generates improvement suggestions. Tools for inventory and production process optimization are on the market. The area of optimization based on operations research is the most advanced and mature area in supply chain management IT supported functions.

- **Sustainability:** This has become a top trend on the agenda of senior executives as the idea that sustainability can have positive effects on the cost side has gained ground. More efficient return and recycle processes, energy-efficient supply chain networks, and waste-avoiding processes have led to significant cost reductions. Technology can help to achieve improvements with sustainable and efficient network design supported by planning and optimization, which enables supply chain planners to define efficient, energy-cost-reducing, and sustainable supply chain structures. Optimization software for supply chain network design that supports green logistics will also gain increasing market share.

Conclusion

There can be no denying the fact that over the past 30 years, the field of supply chain and logistics management has come into its own. There are even commercials on TV that say "I love logistics" (for example, UPS) and mention the importance of a good supply chain (for example, IBM). Thanks to trends such as globalization, outsourcing, the Internet, e-commerce, enterprise resource planning (ERP) systems, and so on, the world has become a smaller place and is much more interconnected.

The outlook for the function and profession is very bright. After reading this book, whether you are interested in the field as a profession or just want to know what it's all about, you should now have a good fundamental understanding of both its operation and importance in the world of business.

References

Chapter 1

Council of Supply Chain Management Professionals (CSCMP) (2014). Retrieved from www.cscmp.org.

Ernst & Young (2014), *Supply Chain Segmentation*. Retrieved from www.ey.com.

Johnson, M. E. (2006). Supply Chain Management: Technology, Globalization, and Policy at a Crossroads, *Interfaces, 36*(3), May–June 2006, 191–193.

Krajewski, L. J., Ritzman, L. P., and Malhotra, M. K. (2013). *Operations Management: Processes and Supply Chains*, 10th edition. Pearson Higher Education, 11–12.

Porter, M. (1985). *Competitive Advantage: Creating and Sustaining Superior Performance*. New York: The Free Press.

Supply Chain Council (SCC) (2014). Retrieved from www.supplychain.org.

Thomas, K. (2012). Supply Chain Segmentation: 10 Steps to Greater Profits. *CSCMP Supply Chain Quarterly*, Quarter 1.

Chapter 2

Dittman, J. P. (2012). *Supply Chain Transformation*. McGraw-Hill.

Hitachi Consulting (2009). Six Key Trends Changing Supply Chain Management Today: Choosing the Optimal Strategy for Your Business, A Knowledge-Driven White Paper.

Lee, H. L. (2004). The Triple-A Supply Chain. *Harvard Business Review, 82*(10), 102–112.

Perez, H. D. (2013). Supply Chain Strategies: Which One Hits the Mark?, *CSCMP Supply Chain Quarterly*, Quarter 1.

Ruamsook, K, and Craighead, C. (2014). A Supply Chain Talent "Perfect Storm?, *Supply Chain Management Review*, January/February 2014, 12–17.

Taylor, V. (2011). Supply Chain Management: The Next Big Thing?, *Bloomberg Business Week*, September 12, 2011. Retrieved from www.businessweek.com.

Chapter 3

Heizer, J., and Render, B (2013). *Operations Management*, 11th Edition. Prentice Hall, 106–107.

KPMG (2007), Forecasting with Confidence, Advisory. Retrieved from www.kpmg.com.

Myerson, P. (2014). *Lean Retail and Wholesale*. McGraw-Hill Professional, 128–132.

Chapter 5

Hayes, R. H., and Wheelwright, S. C. (1979). Link Manufacturing Process and Product Life Cycles, *Harvard Business Review*, January–February, 133–140.

Hayes, R. H., and Wheelwright, S. C. (1979). The Dynamics of Process-Product Life Cycles, *Harvard Business Review*, March-April, 127–136.

Chapter 7

Association of American Railroads (2004). Overview of U.S. Freight Railroads, adapted from Association of American Railroads, July 2004, Overview of U.S. Freight Railroads. Retrieved from www.nationalatlas.gov.

U.S. Department of Transportation, Research and Innovative Technology Administration, Bureau of Transportation Statistics, and U.S. Department of Commerce, U.S. Census Bureau, (2007). Commodity Flow Survey, Retrieved from http://factfinder.census.gov.

Chapter 10

Aberdeen Group (2010). Reverse Logistics: Driving Improved Returns Directly to the Bottom Line, February 2010.

Futin, E. (2010). The Establishment of Global Supply Chain, presented at APEC e-Trade and Supply Chain Management Training Course, November 9, 2010, Hong Kong.

Greve, C., and Davis, J. (2012). Recovering Lost Profits by Improving Reverse Logistics, UPS white paper on reverse logistics, March 19, 2012.

Reverse Logistics Magazine (2009). Best Buy Turning Returns Processing into Profit Center, *Reverse Logistics Magazine*, January 15, 2009. Retrieved from www.reverselogsticstrends.com.

Rogers, D., and Tibben-Lembke, R. (1999). Going Backwards: Reverse Logistics Trends and Practices. Reverse Logistics Executive Council.

Stock, J., Speh, T., and Shear, H. (2006). Managing Product Returns for Competitive Advantage, *MIT Sloan Management Review*, Fall.

Chapter 11

Christopher, M., and Peck, H. (2004). Building The Resilient Supply Chain. *Cranfield School of Management, International Journal of Logistics Management, 15*(2), 1–13.

Heizer, J., and Render, B (2013). *Operations Management*, 11th Edition. Prentice Hall, 438.

Kauffman, R. G., and Crimi, T. A. (2005). A Best-Practice Approach for Development of Global Supply Chains. 90th Annual International Supply Management Conference, May.

PwC and the MIT Forum for Supply Chain Innovation (2013). Making the Right Risk Decisions to Strengthen Operations Performance. Retrieved from www.pwc.com.

SCDigest (2010). The Five Challenges of Today's Global Supply Chains. August 12, 2010. Retrieved from www.scdigest.com.

Sheffield, Y., and Rice Jr., J. B. (2005). A Supply Chain View of the Resilient Enterprise, *MIT Sloan Management Review, 47*(1), 41–48.

Chapter 12

Greaver, M. F. (1999). Strategic Outsourcing. *AMACOM* (an imprint of American Management Association publications). Retrieved from www.asaecenter.org.

Con-way.com (2014). 4PL Supply Chain Transformation, Menlo Worldwide Logistics white paper. Retrieved from www.con-way.com.

Armstrong & Associates, Inc. (2007 and 2013). Ryder Supply Chain Solutions Site Visits - 3PL Case Study Reports. Retrieved from www.3plogistics.com.

Chapter 13

SAP (2007). Supply Chain Collaboration: The Key to Success in a Global Economy. Retrieved from www.sap.com.

Andrews, J. (2008). CPFR: Considering the Options, Advantages and Pitfalls, Plan4Demand Solutions. Retrieved from www.sdcexec.com.

Chapter 14

Bozarth, C., and Handfield, R. (2008). *Introduction to Operations and Supply Chain Management*, 2nd edition. Pearson, 516–518.

Gartner.com (2014). Gartner Says Worldwide Supply Chain Management Software Market Grew 7.1 Percent to Reach $8.3 Billion in 2012. Press release. Retrieved from www.gartner.com.

Gilmore, D. (2010 and 2013). Insight from the 2010 [2013] Gartner Supply Chain Study. *Supply Chain Digest*, June 8, 2010 and June 28, 2013. Retrieved from www.scdigest.com.

Harrington, L. (2007). Defining Technology Trends, *Inbound Logistics*, April.

Intermec Technologies Corporation (2007). Top 10 Supply Chain Technology Trends. White paper. Retrieved from www.intermec.com.

McDonnell, R., Sweeney, E., and Kenny, J. (2004). The Role of Information Technology in the Supply Chain. *Logistics Solutions, 7*(1), 13–16.

Simatupang, T. M., and Sridharan, R. (2001). A Characterisation of Information Sharing in Supply Chains, Massey University, October.

Earpsearch (2014). Supply Chain Management Software. White paper. Retrieved from www.erpsearch.com.

Chapter 15

Heizer, J., and Render, B (2013). *Operations Management*, 11th Edition. Prentice Hall, 329–330.

Chapter 17

Myerson, P. (2012). *Lean Supply Chain & Logistics Management*. McGraw-Hill Professional, 163.

Chapter 18

Myerson, P. (2012). *Lean Supply Chain & Logistics Management*. McGraw-Hill Professional, 11–16.

Chapter 19

Aronow, S., Hofman, D., Burkett, M., Nilles, K., and Romano, J. (2014). The Gartner Supply Chain Top 25 for 2014. Retrieved from www.gartner.com.

CSC (jointly with Neeley Business School at TCU and *Supply Chain Management Review*) (2014). The Ninth Annual Survey of Supply Chain Progress. Retrieved from www.csc.com.

Nabben, H. (2014) 12 Trends That Are Shaping the Future of Logistics. September 16, 2014. Retrieved from www.supplychaindigital.com/logistics.

Schneider-Maul, R. (2014). How Will Digital Impact SCM: Supply Chain Trends. September 9, 2014. Retrieved from www.capgemini-consulting.com.

scmworld.com (2014). The Chief Supply Chain Officer Report 2014, Pulse of the Profession. September. Retrieved from www.scmworld.com.

Index

D

dashboards, 273-274

data analytics, 266-267

data versus information, 211-212

DCs (distribution centers), 128

decline (product lifecycle), 41-42

defects, 280

delivery, 99, 150. *See also* transportation systems

Dell Computer, 10-11, 226

Delphi method, 39

demand and supply risk, 182

demand drivers, 36-37
  external demand drivers, 37
  internal demand drivers, 36

demand options, 76-77

demand planning
  demand management, 215
  forecasting, 37-38
    ABC method, 46-47
    associative models, 40, 44-46
    Delphi method, 39
    demand drivers, 36-37
    forecasting realities, 35-36
    forecasting software, 46-47
    history of, 34-35
    jury of executive opinion, 39
    knowledge of products, 38-39
    metrics, 46-48
    process steps, 37-38
    product lifestyles and, 40-42
    pyramid approach to, 34-35
    quantitative versus qualitative models, 38-40
    time series models, 40, 42-43
    types of forecasts, 36
  overview, 23, 33-34
  technology and best practices, 46-47
  typical planning and scheduling process, 33

demand planning cross-functional meetings, 74

demand time fence (DTF), 80

density
  density rates, 116
  transportation costs and, 112

dependent demand inventory, 51-53

deregulation, effects on pricing, 115

descriptive analytics, 266

design
  distribution network types, 228
    distributor storage with carrier delivery, 230-231
    distributor storage with customer pickup, 233-234
    distributor storage with last-mile delivery, 232-233
    e-business impact, 235-236
    manufacturer storage with direct shipping, 228-229
    manufacturer storage with direct shipping and in-transit merge, 229-230
    retailer storage with customer pickup, 233-234
  facility layout
    assembly line design and balancing, 256-258
    career outlook, 262
    cellular layouts, 253-254
    facility design in service organizations, 255-256
    fixed-position layouts, 255
    hybrid layouts, 253
    overview, 249-250
    process layouts, 250-252
    product layouts, 250
    technology, 261
    warehouse design and layout, 260-261
    work cell staffing and balancing, 258-259

E

earliest due date (EDD), 85

e-business impact on distribution networks, 235-236

Economic Order Quantity (EOQ) inventory model, 55-57

economic value analysis (EVA), 267

economics

transportation

cost elements, 113-114

cost factors, 111-112

shipping patterns, 112

transportation economics, 110-111

warehouses

economic benefits, 131-135

economic needs for warehousing,
126-127

ECR (Efficient Customer Response),
207-208

EDD (earliest due date), 85

EDI (electronic data interchange), 99

Efficient Customer Response (ECR),
207-208

EFT (electronic funds transfer), 99

electronic data interchange (EDI), 99

electronic funds transfer (EFT), 99

emerging supply chain technology trends,
219-221

employees

hiring, 77

laying off, 77

part-time workers, 77

subcontracting, 77

temporary workers, 77

end-to-end visibility, 296

engineer-to-order (ETO), 79

enterprise resource systems (ERP), 216

enterprise solutions, 213

environmental considerations for reverse
logistics, 170-172

environmental risk, 182

EOQ (Economic Order Quantity) inventory model, 55-57

EPR (extended product responsibility)
programs, 171

e-procurement, 99

ERP (enterprise resource systems), 216

error measurement, forecasting, 46-48

mean absolute deviation (MAD), 47

mean absolute percent error (MAPE),
47-48

mean squared error (MSE), 47

tracking signals, 48

ETO (engineer-to-order), 79

EVA (economic value analysis), 267

evaluating vendors, 96

event management, 216

exception rates, 116

execution

execution viewpoint, 214

execution-driven planning solutions,
221

execution-level collaboration, 203

supply chain execution, 215-216

executive opinion, jury of, 39

expediting, 114

exponential smoothing, 43

extended enterprise solutions (XES), 213

extended product responsibility (EPR)
programs, 171

external demand drivers, 37

external integration, 201-206

benefits of, 206

collaboration methods, 206

CPFR (collaborative planning, fore-
casting, and replenishment), 208-210

ECR (Efficient Customer Response),
207-208

QR (quick response), 206-207

supply chain segmentation, 298

supply chain strategy, 15
  elements and drivers, 17-19
  methodology, 19-23
  mission statement, 15-16
  strategic choices, 17
  SWOT analysis, 16-17

supply chain technology, 211
  best of breed solutions, 213
  bullwhip effect, 212
  customer relationship management
    (CRM) software, 214
  data versus information, 211-212
  emerging trends, 298
  enterprise solutions, 213
  information needs, 213-214
  point solutions, 213
  product lifecycle management (PLM)
    software, 214
  supplier relationship management
    (SRM) software, 214
  supply chain information flows, 212
  supply chain software market, 214-218
    *best-in-class versus single integration
      solution, 217*
    BI (business intelligence), 216
    consultants, 218
    costs, 217
    emerging trends, 219-221
    short-term trends, 218
    supply chain event management, 216
    supply chain execution, 215-216
    supply chain planning, 215
  trends in, 297-299
  XES (extended enterprise solutions), 213

supply chain trends, 294-295

supply management, 215

supply planning cross-functional meeting,
  74

supply risk, 182

support, reverse logistics, 162

Surface Transportation Board (STB), 115

surveys, market surveys, 39

sustainability, 171, 295-299

switching and terminal (S&T) carriers,
  107

SWOT analysis, 16-17

system design, reverse logistics, 162-164
  documentation, 165
  product collection system, 164-165
  product location, 164
  recycling or disposal centers, 165

T

tactical collaboration, 202

tactical viewpoint, 213

tactical-level measurement, 268

talent development, 295

TAPS (Trans-Alaska Pipeline System),
  108

teams (Lean), 281-283

technology, 211
  aggregate planning and scheduling, 86
  best of breed solutions, 213
  bullwhip effect, 212
  customer relationship management
    (CRM) software, 214
  data versus information, 211-212
  development, 8
  emerging trends, 298
  enterprise solutions, 213
  facility layout, 261
  information needs, 213-214
  network optimization solutions,
    246-247
  OMSs (order management systems),
    155-156
  point solutions, 213
  procurement, 101
  product lifecycle management (PLM)
    software, 214

layouts, 253-254

staffing and balancing, 258-259

work in progress (WIP), 53

workflow, 6

X-Y-Z

Xerox, 171

XES (extended enterprise solutions), 213

yard management system (YMS), 145

YMS (yard management system), 145

zero returns, 169